GOING GREEN

The Essential Guide

Need
— 2 —
Know

Cora
Lydon

First published in Great Britain in 2010 by
Need2Know
Remus House
Coltsfoot Drive
Peterborough
PE2 9JX
Telephone 01733 898103
Fax 01733 313524
www.need2knowbooks.co.uk

Need2Know is an imprint of Forward Press Ltd.
www.forwardpress.co.uk
SB ISBN 978-1-86144-089-1
Cover photograph: Stockxpert

Contents

Introduction

Did you know that the volume of waste produced in just one hour by UK residents would fill the Albert Hall? Or that the average woman comes into contact with over 515 chemical compounds every day? With such alarming statistics, it's no surprise that everyone is trying to do their bit for the environment. But leave aside thoughts of cancelling your holiday, surviving on mung beans or wearing hemp sacks, going green is easier than you think and this book will help you do it in style.

The last decade has seen a major turnaround on how we live our lives – pop stars line up to sell songs for the environment, our country's leaders have put climate change at the top of the agenda and major corporations are planting forests and spearheading the green movement. But what can one solitary person do? Well, more than you might think.

From homeowners and fashionistas to children and those on a tight budget, greening over your life can be fun, inspirational and, more often than not, it will also save you money too. It's time to stop thinking that there's little point in ditching plastic carrier bags because you're just one person – just one person can make a difference as every little bit helps to conserve the Earth for future generations. With the help of this book you can learn how to make both small and big changes which will suit your lifestyle and help to protect the environment.

'It is the greatest of all mistakes to do nothing because you can only do a little – do what you can,' advised the Reverend Sydney Smith, and some 150 years later the sentiment is just as relevant today as it was then. That sentence perfectly sums up the ethos of this book. It will be encouraging you every step of the way to make small changes that will fit in with your lifestyle. It won't be preaching – going green isn't, and shouldn't be, a chore. When it comes to the environment, every little bit really does help – so no matter how big or small the changes are that you make, you can be confident that they really do count.

> 'It is the greatest of all mistakes to do nothing because you can only do a little – do what you can.'
> Reverend Sydney Smith.

One person making small changes is all it takes to get people going green, and as you begin to see the other benefits of being more environmentally-sound (such as improved health and saving money), you'll be keen to extol the benefits to your friends and loved ones who in turn could follow your lead. Enthusiasm is catching after all!

Enjoy your journey to a greener lifestyle.

Disclaimer

The content of this book is for guidance only and no guarantees can be made for potential energy or cost savings. The publisher and author accept no responsibility for damage to possessions when following green tips listed throughout this book. For the most up-to-date guidance on green issues, consult the websites of organisations listed in the help list.

This book has been printed on FSC certified paper. Our books are printed using the latest Océ digital technology, designed to help keep carbon emissions to a minimum.

Chapter One

Why Go Green?

Going green has become such an ingrained part of today's lifestyles – the message is everywhere: make changes to the way you live your life to prevent further damage to our environment. But, the advice can be confusing, difficult to follow, and often fails to explain just how beneficial your actions could be. The aim of this book is to help you reduce your carbon footprint through simple actions. The UK has a total carbon footprint of around 650 million tonnes, which means the average carbon footprint of an individual stands at approximately 11 tonnes per year.

What are the benefits of going green?

By now we're all well aware of what going green means for the environment – but what other benefits can you expect to see? The first is your bank balance. Going green is all about reconsidering your purchases – do you need to replace an item of furniture or do you have something up in the loft that could be given a makeover? It's also about getting the most from the energy you use – for example, heat proofing your home so that valuable heat can't escape or being resourceful with your water.

Being more environmentally aware can also have a big impact on your family. Growing your own veg, cycling to the shops together and working to conserve the local environment are all things that can be enjoyed together, helping you spend quality time forging strong bonds as a family and ensuring that your children grow up to be responsible citizens.

'The UK has a total carbon footprint of around 650 million tonnes, which means the average carbon footprint of individuals stands at approximately 11 tonnes per year.'

Health warning

The health and fitness of you and your family will also be drastically improved by cutting out harmful chemicals and leaving the car at home. On a larger scale, decreasing the pollution in our atmosphere could lead you to live a longer life. According to a report by the cross-party Environmental Audit Committee, air pollution could be wiping up to nine years off our lives. Evidence collected suggests that the problem may be contributing to as many as 4,000 deaths every month in the UK, as it increases the risk of several conditions including asthma attacks, cancer and heart disease.

Finally, going green is all about making you feel good for the changes you're making. Knowing that every action you change works towards preventing one more tree from being felled, extending someone's life by improving air quality or protecting habitats for wildlife should help to put a smile on your face.

Take action

As you read this book, be inspired to take action. All of the ideas are designed to be implemented easily into your current lifestyle, so why not try picking out a new behaviour or action to make every week? There's lots of practical advice here, and within the help list you'll find details of websites that can provide you with even more information for overhauling your lifestyle.

Climate change Q&A

Q. What is your carbon footprint?

A. Many of the actions you take leave behind a stain on the environment. If you're relying on something other than your own energy for power – such as petrol for the car, fuel for the cooker, gas to heat the home – you can be fairly sure it's adding to your carbon footprint. Most of the energy that we use is created by burning fossil fuels like gas, coal and oil. As these burn they release greenhouse gases into the atmosphere, increasing global warming. Our carbon footprints are calculated based on the amount of greenhouse gases we're responsible for emitting.

Q. What are greenhouse gases?

A. While some greenhouse gases are emitted naturally, e.g. water vapour, carbon dioxide (CO_2), ozone, methane and nitrous oxide, many are released as a result of human action. When we burn solid waste, fossil fuels or wood products, CO_2 is released. Nitrous oxide gases are commonly emitted during agricultural and industrial processes, while methane is released as organic waste breaks down – in both livestock farming situations and landfill sites.

These greenhouse gases trap heat in the Earth's atmosphere and by raising the Earth's temperature they ensure our planet is warm enough for humans to live on. Naturally occurring greenhouse gases do an excellent job of keeping our Earth at a comfortable temperature, but the added burden of man-made greenhouse gases is becoming catastrophic. If the greenhouse gases become too strong, they'll cause the Earth to heat up more than necessary – as we have seen in the last 150 years.

Q. What does global warming mean?

A. Global warming refers to the increasing temperature of our planet and scientists are concerned about the effects this will have for future generations. Global warming doesn't just mean the Earth will get hotter; this increased heat also means rising sea levels as ice caps melt – which in the long term could leave low lying areas underwater – and extreme weather conditions which could lead to droughts, heavy rainfall and even tropical cyclone activity.

Global warming will lead to climate change – that is, changes in our climate that will have a dramatic effect on our health and wellbeing. The impact of increased CO_2 in the atmosphere, higher temperatures and changes in the frequency and volume of rain will alter the world's agriculture and food production chains. When it comes to our health, the higher temperatures may lead to disease-infected insects migrating and spreading diseases. Not to mention the toll extreme weather conditions could take on our safety.

The most comprehensive report into the impact of climate change in the UK was published in June 2009. The United Kingdom Climate Projections expect that our winters will become warmer and wetter and our summers hotter and drier – plus we may face drought, flooding, rising sea levels and intense heatwaves unless we act now.

Q. Is it too late to do something?

A. It's never too late to do something to help preserve our environment. If we do nothing we're pretty much resigning ourselves to the fact that our Earth won't be habitable for future generations. But if we all start to make changes and encourage friends and family to follow our lead, over time we can help to secure a happier, greener life for future generations. This involves protecting the Earth's vast natural resources, slashing the volume of waste we create and reducing our carbon footprints. Scientists predict that global greenhouse gas emissions need to peak in the next decade and then fall well below their current levels in order to prevent dangerous changes to our environment.

Q. What is greenwashing?

A. As the environment has hit the headlines, companies are falling over themselves to declare themselves as environmentally sound, knowing that given a choice people are likely to opt for the most ethical company if all other variables remain the same. In short, greenwashing is the disingenuous spin that some companies will put on its products, services or processes in order to gain new consumers. This can make it hard for people to spot a truly green item – if you're at all unsure, consider the following points below before making a purchasing decision:

- Can the company prove they have eco credentials that are accredited to a specific scheme or can they at least substantiate their green claims?

- Look out for how integrated their green practices are – a company will need to do more than just minimise its waste. True green practices should touch all areas of the organisation.

- Don't be afraid to ask questions of the company. If they fail to give you a satisfactory answer then reconsider doing business with them. Any company should be happy to share their green success and be able to talk in depth about what this means.

- Evaluate a product or service at all levels – does your parcel arrive swaddled in layers and layers of packaging and in a box that is far bigger than necessary? Does the service engineer arrive by public transport, in a hybrid vehicle or in a CO_2-spewing car?

Green facts

If you're still ensure of the importance of going green then take a look at these facts to spur you into action:

Just 1% of Australia's untapped geothermal power potential could hold enough energy to last for a staggering 26,000 years.

According to Pilkington energiKare in March 2010, 76% of homebuyers said energy efficiency was a deciding factor when purchasing a home – people were most concerned with loft insulation, energy efficient boilers, heating, windows and cavity insulation.

Since 1850, when reliable temperature data was first recorded, the 10 warmest years on record have all taken place since 1997. Records indicate that spring arrives an average of 10 days earlier than it used to in the 1970s.

If we experience just two more degrees of global warming, scientists are predicting that around 20% of the Earth's species will become extinct.

The average plastic bag takes up to 500 years to decompose – and most people admit to only using the bag once.

In just one day the UK produces enough waste to fill Trafalgar Square up to the top of Nelson's Column. In one year it will fill enough dustbins with waste to stretch from the Earth to the Moon.

'The average plastic bag takes up to 500 years to decompose – and most people admit to only using the bag once.'

Summing Up

It's time to stop thinking that it's too late to stop the devastation of the planet and start to do our bit to ensure its survival. We've all contributed to the environmental destruction of Earth and now it's time for us all to take action. It doesn't require vast sums of money – you just need to understand the impact of your actions and look for alternative ways to carry them out.

Say no to plastic bags at the supermarket and take your own reusable bags, sort your waste ready for recycling before you bin it, switch lights off before you leave a room – simple steps just require a change in your mindset but can have a dramatic impact on the Earth's future.

We all have a responsibility to make these changes, and by working together with your friends and family you really can make a difference. Plus you'll also experience the side effects of going green too – better health, more money in your pocket and stronger family links.

The more you delve into the topic, the more you'll be able to spot the company looking to boost its profits rather than lower its environmental impact. Be empowered to ask questions so you can spot greenwashing a mile off and take a stand against it.

Chapter Two

How Big is Your Carbon Footprint?

When it comes to determining your carbon footprint, your first port of call should be the Internet. At the Act on CO_2 website (see help list), there's a very useful carbon footprint calculator which will estimate your carbon footprint by asking a series of questions about appliances in your home, how you travel and your energy usage. According to Act on CO_2 the national average household emits 10.17 tonnes of carbon every year – which is enough to fill more than 344,212 party balloons or boil 548,991 cups of tea!

So, not sure whether your carbon footprint is Godzilla-sized or more dainty? Take a look at the following statements to get a rough guide on how you're doing so far and to identify the areas you could look to improve upon.

Home

Is your home:

- Pre 1920s?
- Pre 1990s?
- Brand spanking new?

Although it varies from property to property, some homes can actually help reduce your carbon footprint. A report from the Empty Homes Agency indicates that the carbon released when building a new home accounts for nearly three times as many emissions as are released over the building's lifetime. However, some new homes are packed with eco-initiatives to help keep your footprint down, though very large homes will require a lot more energy to keep them warm. Moving to a pre-loved home will keep your carbon emissions low but look for one that is still fairly modern. Older homes emit an average of eight tonnes of CO_2 per year, whereas more modern properties can reduce that by nearly half.

Is your boiler:

- More than 10 years old?
- Under 10 years old?
- I don't have a boiler.

In a gas-heated home, the boiler accounts for around 60% of the CO_2 emissions. The key is to ensure your boiler is as energy efficient as possible – a G rated boiler (G is the worst rating for boilers, while A-rated is the most energy efficient) will really push up your personal footprint compared to a high efficiency condensing boiler. A high efficiency boiler captures much more useable heat from the fuel source compared to the non-condensing kind. If you have an old boiler, think about replacing it with a new model – boilers are now labelled according to how environmentally friendly they are, so always check the energy efficiency label before you buy. See page 91 for more information.

Do you have:

- Double glazing?
- Cavity wall insulation?
- Loft insulation?

All of the above options will help you save money on your fuel bills and bring down your carbon footprint. Around a third of the heat in your home is lost through the walls, but by insulating them you can really improve on this figure. Solid wall insulation is also possible too. Loft insulation is a clever way to trap in heat and knock off hundreds of pounds from your energy bills. Draught proofing your home – through sealing leaky window and doorframes, filling in gaps in floorboards and using draught excluders – will also make a difference. Finally, double glazing works by trapping air between two panels of glass to form an insulating barrier – keeping you toasty warm for less.

Travel

Do you use your car:

- For every journey?
- For every long journey?
- Only when necessary?

All of those short car journeys soon add up to one big problem – your CO_2 footprint. Walking or cycling for shorter journeys means you'll get fit, save money and knock that carbon footprint into shape. For longer journeys, don't automatically turn to your car – trains and coaches can be a more environmentally-friendly way to travel. Or perhaps you can team up with other people heading in the same direction as yourself and just take one car between you.

How often do you get your car serviced?

- Yearly.
- Rarely.
- Never.

A well-looked-after car is a more energy efficient car, so having your motor serviced yearly will keep on top of any niggles which could be blowing your eco halo to shreds. For maximum fuel efficiency, tyres need to be pumped up to their optimum levels – so remember to check them regularly!

'Walking or cycling for shorter journeys means you'll get fit, save money and knock that carbon footprint into shape.'

Appliances

Do you use your washing machine on:

- 30 degrees?
- 40 degrees?
- 60 degrees?

Studies show that laundering accounts for between 60 and 80% of a garment's total environmental impact and so it's a wise idea to get to know the most energy efficient way to clean your clothes. A 60 degree wash will use almost half as much energy as a 90 degree wash, but a 40 degree wash will use almost half as much as a 60 degree wash. And, according to washing powder manufacturer Ariel, over the course of a year we could save enough CO_2 to fill four million double-decker buses if we all switched from a 60 degree wash to a 30 degree wash. You can also further shrink your CO_2 footprint by around 45kg by only doing full loads.

'Tumble driers are the most energy hungry of all domestic appliances, typically emitting between 330kg to over 800kg each year.'

Do you:

- Line-dry your washing?
- Hang it on airers in the home?
- Use your tumble dryer?

Solar power is not just good for powering your devices – it's also efficient at drying your laundry too! Whenever you can, dry washing out on the line – even an overcast day can be warm or windy enough to dry clothes. In the depths of winter, use airers to hang washing out to dry, and use your tumble dryer only when you really need to. Tumble driers are the most energy hungry of all domestic appliances, typically emitting between 330kg to over 800kg each year. So use them sparingly. However, avoid draping clothes over radiators as this will cause your boiler to have to work even harder, and don't think about turning up the heating to dry your clothes – in this situation it would be far better to run the tumble dryer than to increase the heat in the whole household.

Where is your phone charger?

- On the window ledge – it runs off solar energy.

- In a drawer in the kitchen.

- Plugged in (and switched on).

You might think one tiny charger can do little damage, but you'd be wrong. The cost of leaving that phone charger plugged in and switched on all the time could soon add up, especially when combined with leaving other appliances on standby. According to the Energy Saving Trust, leaving chargers switched on, lights on and appliances on standby could be pumping an extra 43 million tonnes of CO_2 into the atmosphere. So the general message needs to be – if you're not using it, switch it off completely.

Summing Up

The first step towards reducing the burden you place on the environment is understanding how the actions that you make on a daily basis can be catastrophic for the Earth. Simple things that we take for granted – switching on the heating when we feel the first chill, making a cup of coffee, jumping in the car – add up; all of the power has to come from somewhere and understandably it leaves behind a footprint.

It's important to assess what areas of your life could do with an overhaul to make them more environmentally friendly. As we've said before, every little helps – so if you want to start small, just tackle one area at a time and see how easy it is. Before you know it, those green steps will become second nature.

Chapter Three

Going Green at Home

Our homes are the area we are most likely to implement green initiatives in, but that's not to say there isn't more that could be done. The Energy Saving Trust has a free home energy check which will tell you how you can save both money and wasted energy. Visit their website (listed in the help list) for more information on this. Paula Owen, an energy doctor at the Energy Saving Trust says: 'We spend around £1,200 a year on fuel to power our homes, and each UK home emits almost twice as much CO_2 as the average car emits in a year, every year. There are lots of ways to significantly reduce a home's emissions and running costs, starting with your daily habits.'

Every little counts

The changes you make to your home can be as radical or as simple as you want. If you want to start small then think about replacing all of your light bulbs with energy saving versions, which will save you approximately £60 over the course of the light bulb's life. Simple measures, such as drawing your curtains as soon as dusk sets in and swapping a bath for a shower, are easy but free steps you can take right now. 'Small efforts can save cash,' confirms Paula. 'For example, how you make a cup of tea – if we all boiled only the water needed when we used the kettle, it'd save enough electricity in a year to run the UK's street lighting for seven months; or turning your thermostat down – reducing room temperature by 1°C can cut heating bills by up to 10% and save around £55 per year.'

There are some steps you can take which do involve an initial outlay – but often they will increase your cost savings too, so that over a period of time they do pay for themselves. Paula says: 'Next you can evaluate the fabric of your home. Quick DIY jobs, such as draught proofing, can shave around £25 a year off heating bills (if everyone in the UK draught-proofed their homes, collectively we'd shave almost £200 million off the nation's annual domestic fuel bill). Other projects, such as loft or cavity wall insulation, can be completed over a weekend. Insulating an un-insulated loft could shave around £150 per year off your energy bills (if everyone in the UK topped-up their loft insulation to 270mm, collectively it would shave £520 million off the nation's annual domestic energy bill).' If you have some money to spare then do consider steps such as loft insulation, double glazing or replacing old, energy-inefficient appliances.

Recycling

'When we recycle, used materials are converted into new products, reducing the need to consume natural resources.'

Laura Underwood, Recycle Now.

Recycling is the obvious area which all households can get involved with. The latest figures compiled by the Department for Environment, Food and Rural Affairs (DEFRA) show an average household recycling rate of 36.9% over the last three months of 2008. The same period for the year before had a recycling rate of 33%, so it's clear that Britain's households are getting the message, although there's still a lot more that can be done. 'When we recycle, used materials are converted into new products, reducing the need to consume natural resources,' explains Laura Underwood from Recycle Now. 'If used materials are not recycled, new products are made by extracting fresh, raw material from the Earth, through mining and forestry.'

Get to know your labels

Most of the packaging you come across in your home will include some kind of logo to inform you whether the packaging is recyclable or not. Since being launched in 2009, the British Retail Consortium's universal, on-pack recycling label has been gaining momentum. As well as many grocery retailers, a significant proportion of non-food retailers have also signed up to the scheme – which currently includes Heinz, Duchy Originals, PepsiCo and John Lewis Partnership. The symbol has been designed to inform consumers how likely it is that a particular piece of packaging can be recycled, based on the UK's local authorities' recycling schemes.

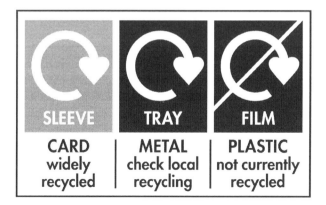

- Widely recycled – indicates that at least 65% of local authorities will collect this type of packaging.

- Check local recycling – between 15-65% of local authorities will collect this form of packaging, so it's worth checking with your own.

- Not currently recycled – less than 15% of local authorities will collect this type of packaging for recycling.

Divide and conquer

You can make the task of recycling a bit easier by dividing up your waste as soon as you bin it. As well as your regular bin for non-recyclable waste, consider keeping a box for recyclable items (think paper, junk mail, cereal boxes, yoghurt pots) and another box for glass – which may or may not be collected alongside your recycling collection. Not all local authorities will collect glass – if yours doesn't, find out where your nearest collection point is. Many supermarkets have bottle banks, so you could get into the habit of taking all your glass bottles and jars on your weekly shop and disposing of them that way.

Can I recycle...?

The Recycle Now website (www.recyclenow.com) has an extensive database of what you can and can't recycle. To get you started:

Yes

- Yellow Pages – 99% of local authorities now accept Yellow Pages for recycling. Or you could check out the company's own recycling scheme at www.yellgroup.com.

- Batteries – not all local authorities will accept batteries in a kerbside collection, so you may need to go to a recycling centre. However, shops selling more than 32kg of batteries a year are now obliged to provide recycling collection facilities in-store. This includes supermarkets – so if you live near any of the big chains, it's easier than ever to recycle your batteries.

No

- Certain light bulbs – another good reason to switch to greener lighting. The old-style incandescent light bulbs cannot be recycled, but your energy efficient bulbs can be recycled at most points.

- Paint – unwanted paint can't be recycled but it can be reused. Community Repaint operates a network of paint donation schemes where you can give unfinished paint to benefit community and voluntary groups.

Check

- Shredded paper – some local authorities will not accept shredded paper, as not all paper mills will accept it for recycling.

- Wrapping paper – there are several reasons why your wrapping paper may not be accepted by your local authority for recycling: for example, it is often covered in sticky tape, may be of poor quality and so has few recyclable fibres, and certain finishes – such as glitter, plastics or non-paper additives – can't be recycled.

Clean sweep

The number of chemicals in the home is rising and our obsessive cleaning methods are largely to blame. Today's average homeowner wants something that requires as little elbow grease as possible but delivers sparkling results, and unfortunately all too often this means chemicals. However, slowly but surely, green cleaning products have become available and deliver outstanding results without the health warnings.

In the last decade or so, levels of asthma have been on the increase – in men we are now seeing a 29% increase in 10 years, while in women it is a staggering 82%. In part, experts believe this is down to larger fat reserves in the female body which is where toxins are stored, but as the majority of women are also the domestic element in the home, and so exposed to harmful cleaning products, our increasingly toxic lifestyles could also be to blame.

Green cleaning tricks

As well as a great range of eco-friendly products available in the shops, there are also many natural ingredients which can be used at home:

- Vinegar and hot water is an excellent treatment for dealing with greasy surfaces, including glassware and windows. Once you've washed those windows, buff them with newspaper to get them to shine.

- If your oven could do with a thoroughly good clean, mix up a thick paste of bicarbonate of soda and water. This can be used in the oven and on the top of the cooker too.

- Tackling limescale doesn't require a harsh cleaner. Soak some cotton wool in neat malt vinegar and wrap it around affected taps and fittings, cover with a plastic bag and secure. After an hour or two remove, rinse and tackle any stubborn remains with a cuticle stick.

- Switch your polish for beeswax – it's a totally natural product and though it does require a little bit more polishing power, the results are excellent. The best technique is to apply a thin layer of wax, allow it to harden for a short while, then burnish with a soft, clean cloth, rubbing in the direction of the grain.

'If your carpets could do with a little tender loving care, sprinkle them with bicarbonate of soda, leave for several hours and then vacuum up.'

- If your carpets could do with a little tender loving care, sprinkle them with bicarbonate of soda, leave for several hours and then vacuum up.

- To deal with smelly drains, pour half a cup of baking soda down the drain, follow with a cup of vinegar and allow the mixture to foam for a few minutes. Finish with some boiling water for clean and fresh drains.

- Lemons have a mild bleaching action as well as antibacterial properties. Try rubbing half a lemon on fabric stains and leave to absorb for half an hour, then wash off (always test the fabric first on an unobtrusive corner).

- Half a lemon popped in the refrigerator will help to eliminate nasty odours.

- Olive oil is perfect to remove stubborn stickers from glass or plastic surfaces – just apply some oil to the area, leave for 10 minutes so it can penetrate and then gently scrub off.

- If your bathroom is prone to mildew, tackle it with a few drops of tea tree essential oil diluted in a spray bottle of water, then use to clean the bath and shower.

What to buy

At the supermarket, look out for specific green cleaners which will be free of toxins and chemical ingredients. Some brands to look out for include: Ecover (www.its-ecological.com), Natural House by Bentley Organic (www.natural-house.co.uk), E-cloths (www.e-cloth.com), Bio-D (www.biodegradable.biz) and Method (www.methodproducts.co.uk).

Home hazards

If you're looking to buy new furniture or redecorate your home then you should familiarise yourself with some of the worst home hazards to avoid. These ingredients are found in many of the products we furnish our homes with but can be detrimental to our health. As many are chemical-based, they are no friend to the environment either.

Home hazard: brominated flame retardants (BFRs)

BFRs are used in a range of products in our homes to increase resistance to fire – TVs, computers, washing machines and refrigerators are all likely culprits. There are more than 50 BFRs in common use today but only two have been regulated. As of 2006, their use in electrical and electronic equipment has been prohibited – but you're still at risk if your equipment was purchased prior to this.

Home hazard: phthalates

Phthalates are commonly seen in PVC products, which include vinyl floor tiles and shower curtains, as well as solvents in toiletries. Phthalates are a group of chemicals that are used in large amounts in soft plastic. When you refit your bathroom, consider replacing your shower curtain with a glass shower screen to reduce exposure.

Home hazard: triclosan

This anti-bacterial chemical also goes under the name 'microban' and is added to a range of products such as washing-up liquids, mouthwashes, dishcloths and chopping boards. The fact that it has been detected as a contaminant in human breast milk and fish demonstrates its poor ability to break down in the environment and the strength of its contamination in our bodies. Instead of buying a new chopping board, keep your existing one hygienic by rubbing it with a cut lemon on a daily basis.

Home hazard: formaldehyde

This particular chemical is the most commonly found chemical in the indoors environment and there are many sources including your towels, as it is used as an anti-crease agent in textiles, and in laminated furniture. Look out instead for organic towels made from more natural fibres such as bamboo, which is proven to be highly absorbent.

Home hazard: paint

Conventional paints are made from raw ingredients synthesised from petro-chemicals. The process requires significant amounts of energy and so produces large amounts of greenhouse gas emissions. Look for paint ranges such as Ecos Organic Paints, Earthborn Paints, Pots of Paint and Ecolibrium Paints.

Energy efficiency

Green energy

Burning gas and coal in power stations or getting electricity from nuclear plants has a large impact on the environment. So obtaining energy from a supply that is renewable makes sense. The main renewable energy sources are solar (from the sun), hydropower (energy from water), geothermal (sourced from hot water springs and magma), wind and biomass (from firewood, crop manure and waste).

Launched at the start of 2010, a new certification scheme will now ensure that you can pick a green electricity tariff with confidence. Launched by the energy regulator Ofgem, the Green Energy Certified scheme label will only be available to suppliers who can demonstrate that its tariff will result in a reduction of the minimum threshold of CO_2 emissions. If you're looking to move to a green electricity supplier, Green Electricity Marketplace assesses what's currently available on the market and is a good starting point.

By turning down your thermostat just one degree, you can save yourself around 10% on your annual heating bill and of course lower your carbon footprint. If radiators have individual controls, adjust them so you're not unnecessarily heating empty rooms. Never leave appliances on standby either, as they'll still be using power, and always switch off lights when you leave the room.

Heat loss in an un-insulated home

According to the Energy Saving Trust, if your home doesn't have insulation you'll be losing significant amounts of heat:

- 33% lost via walls.
- 26% lost via roof.
- 18% lost via windows.
- 12% lost via drafts.
- 11% lost via floorboards and doors.

How does your garden grow?

When you're focusing on being as green as possible at home, don't forget to consider your garden too. There are many ways to ensure your garden has as little impact on the environment as possible. One of the first things to look at is the use of water – and how you can recycle it for use in the garden. Fitting a water butt will collect rainwater that can then be used to water your garden with.

You can also ensure that any water used in the home works twice as hard through a grey water system. This effectively means reusing water from your bathtub or washing machine on the garden. There are various systems available to make the process much easier, or you could start by using buckets to scoop out water from the bathtub to use in the garden or tipping washing-up water out onto the garden. If you're doing this it's always best to ensure you're using chemical-free products in the water to avoid unnecessary pollution of your garden. When using a hosepipe in the garden, always fit a trigger to control the flow of water and where possible opt for plants that need minimal watering.

'By turning down your thermostat just one degree, you can save yourself around 10% on your annual heating bill and of course lower your carbon footprint.'

Go peat-free

It's also a wise idea to stop using peat in the garden. It takes many centuries for a peat bog to form, yet modern machinery can destroy it in just a matter of days and in doing so also devastates precious wildlife habitats. Not only that, but an estimated half a million tonnes of CO_2 is emitted every year during the peat extraction process. Peat-free composts made from wood bark, wood fibre and green waste are available, so look out for them or create your own (see chapter 5).

Summing Up

Our home is the place where we spend most of our time and so it makes sense to start your journey on a greener path here. It's easy to get the whole family involved in eco initiatives and you can start off small. Simple changes to your lifestyle, such as changing behaviours and rethinking your waste, soon add up to a brighter, greener future.

Our homes are also one of the most energy unfriendly areas, particularly when loaded with modern technology. Most homes today have at least one TV, DVD player, computer and games console, which soon adds up to one unhealthy carbon footprint. Start by getting everyone to switch off appliances when they're not in use or turning off lights as they leave a room. Don't forget it's not just about the power your home uses – water is an equally valuable resource and can be reused around the home and garden for maximum benefit.

Once these small steps are in place, you could look at what bigger changes you'd like to make. Loft insulation, cavity wall insulation, double glazing and replacing your appliances with more eco-friendly models can all help reduce carbon emissions. Although these require upfront costs, they certainly pay off in the long term. If you're struggling to convince other members of your family to follow your lead, try showing them some of the cost savings or startling environmental facts to sway them.

Chapter Four

Going Green at Work

While many people have started to make simple changes at home, green lethargy is still rife in offices up and down Britain. James Strawbridge, eco expert, co-presenter of the BBC series *It's Not Easy Being Green* and supporter of the Green Office Week campaign, agrees: 'I think at the moment a lot of people are struggling to convert how green they are at home to how green they are in the workplace. They might recycle at home and be conscious of their waste, but as soon as they enter the workplace environment they're not sure how to proceed with environmental practice.'

Nicky Amos, founder and director of Nicky Amos CSR Services Ltd, agrees: 'Businesses are responsible for 40% of the UK's carbon emissions, compared with 27% from households. Most of our individual effort goes into reducing our impacts at home. So why is it that when we leave our homes each morning, the vast majority of us spend our days in over-heated, over-ventilated and over-lit workplaces that consume massive amounts of resources and contribute millions of tonnes of CO_2 to the UK's carbon footprint? The irony is that the efforts we make at home can be just as easily applied to the workplace.'

Why are businesses dragging their feet?

A dangerous combination of presuming one of your colleagues will do it, together with the perceived idea that greening over a business will cost money, all add up to the fact that too many businesses are failing to go green. The fact is, many of the actions we undertake at home are easily implemented in the office too, but a joint survey by Tickbox.net and Opinion Matters in 2009 revealed easy doesn't always make it happen. Over 40% of people who were surveyed confessed that although they recycled plastics at home, they failed to do so in the office. And it's not just recycling that is a missed opportunity

– 33% of survey respondents were guilty of leaving electrical items on in the workplace when they left the office, despite the fact that they'd never do this at home.

Profit from green changes

The fact is that going green can actually be a profitable exercise for companies. As the government introduces more and more environmental legislation, businesses may start to be penalised for not having sound ethical and eco-friendly policies in place. Green Office Week, which launched in 2009, encourages Britain's offices to make green changes. The campaign also highlights the vital importance that office workers place on working in an environmentally friendly office. A YouGov survey confirms that an overwhelming 92% of office workers think it's important that British companies are environmentally responsible.

Taking action

The government is looking ever more closely at tackling environmental issues in relation to offices and by making changes now you can ensure good ethical practice before legislation comes in. From clean air acts to rules on disposal of toxic waste and landfill taxes through to penalties for failure to recycle, one of the fastest growing areas of UK legislation is environmental law. For example, standard rates of landfill tax in the UK are set to increase by £8 per tonne every year until at least 2013, in order to discourage the use of landfill sites. 'It's a good idea to get green disciplines in place while the government has not got taxation in place,' agrees Matt Roper, founder and CEO of GreenBuying.co.uk. 'It's only a matter of time before the government will start targeting companies for carbon tax and you'll be in a better position when this is introduced if you're already acting in an ethical manner.'

Matt takes a refreshing approach to going green – it's not about transforming every action, it's about making every little change happen. 'Some companies still need to do some work to be eco-friendly, but the important thing is that they have a strategy in place to do so. You don't need to aim for perfection – but at the end of the day the message is don't sit around doing nothing. Start incrementally to make changes and over time it will happen.'

Green buying

A key part of this strategy should be looking closely at the purchases your company is making. Before purchasing anything, your first question should be: 'Do we need to make this purchase?' Ask yourself if there is anything in the company that could be recycled. According to Matt, everything you purchase has an impact on the environment – through the use of the raw materials, the energy that is needed to produce it, the packaging, the use of fuel to ship it, etc. So, not buying something is the most environmentally friendly decision you could make. A common practice is that of cascading technology from demanding customer-facing areas of a business through to less demanding areas of an organisation. In practice, this could mean front of house positions or staff working closely with customers are most likely to need new technology, so when employees in these positions are given new computers their old computers could be redistributed to another area of the business which doesn't rely on technology so heavily. This will ensure that computers that still work are given a new lease of life – but in an area that doesn't require the same level of use as previously.

If you do need to purchase something then bear in mind a few tips:

- Never stockpile items – although it may be cheaper to bulk buy, some items deteriorate over time and will end up in the bin. Paper, for example, naturally absorbs moisture, making the product marginally thicker and stickier. Combine this with poor storage that can lead to battered, crumpled or damaged sheets, and stockpiling doesn't seem like such a good idea.

- When purchasing paper for the office, look out for the FSC (Forest Stewardship Council) stamp of approval. This will tell you that the paper has been sourced from sustainable sources and so its impact on the environment will be reduced.

'When purchasing paper for the office, look out for the FSC (Forest Stewardship Council) stamp of approval. This will tell you that the paper has been sourced from sustainable sources and so its impact on the environment will be reduced.'

- When making purchases, always find out how eco-friendly your supplier is – ask questions about their eco policies. 'Your company's carbon footprint is part of a bigger carbon footprint,' explains Matt.

- Look for companies that are ISO 14001 accredited – according to the British Assessment Bureau, this environmental standard means a company is better able to manage their environmental risks and prove their legal and regulatory compliance.

- Think locally, not globally – during the procurement process, find out where the various parts of your purchase have been sourced. To reduce a burgeoning carbon footprint, you want materials that are UK-sourced. In addition, where possible use local suppliers, such as caterers or stationers, to reduce the number of miles your product will have to travel to reach you.

Waste reduction

In just one year, the average worker bins up to a quarter of a tonne of paper – now multiply that figure by the number of people in your office and you'll begin to see just how wasteful your company is. Try to encourage workers to get out of the mindset of printing everything out. Look into online back ups which will reduce the need for filing endless reams of paper. Start a scrap paper box for misprints to be used for people to jot notes down on. Also reset your printer's preference so it always prints on both sides of the paper – an instant way to slash paper usage in half. You can also double the life span of your toner cartridge with a few adjustments. Change the printer to an economy setting for print outs being used internally to get twice as many print outs than a higher quality setting. And if something doesn't need to be in colour, switch to black and white printing.

Recycle now

Of course your office has a recycling bin, doesn't it? But have you ever monitored how many people are religiously using it? According to the Tickbox. net and Opinion Matters joint survey, nearly a quarter of employees (22%) put recyclable items into a non-recycling bin on a regular basis. Many people would cite laziness for this, so make sure recycling bins are scattered around the office and in some of the more obvious places such as by the photocopier, printer and fax.

The way to complete the recycling chain is by purchasing products that have been made out of recycled material. In the space of just a month, that pile of junk mail you binned could land back on your desk as recycled envelopes or notebooks. 'Using recycled materials in the manufacturing process uses considerably less energy than that required for producing new products from raw materials – even when comparing all associated costs including transport, etc,' says Laura Underwood from Recycle Now. 'Plus there are extra energy savings because more energy is required to extract, refine, transport and process raw materials ready for industry compared with providing industry-ready materials.'

Saving energy

Office workers should never ever underestimate just how important switching off office equipment is. If just 50 people switched off their computers and printers every evening, more than 40 tonnes of carbon emissions would be saved – that's the equivalent of taking 11 cars off the road. The average office is a hotbed of electrical equipment – printers, photocopiers, mobile phone chargers, lighting, computers, fans, radios – and if they're not switched off properly at the end of the working day, your company could be using up just as much energy as they do when the office is open.

It's a simple habit to get into, so start making an effort in the office to turn off office equipment. It's a common idea that computers use up more energy to power up from off, than they do by being left on stand-by overnight. But it's a myth – you wouldn't leave your tap running to save yourself the effort of turning it on each morning, so take the same attitude with electrical equipment. While this book is all about empowering the individual to make small changes, when it comes to the office it's also worth communicating these simple actions to your colleagues.

'If just 50 people switched off their computers and printers every evening, more than 40 tonnes of carbon emissions would be saved.'

Light and easy

When it comes to lighting, think before you automatically switch on the lights. By rearranging the furniture in the office, you could make the most of the natural daylight. Not only is this natural daylight healthier for workers, but it will also save a significant amount on the company's electricity bill. Move desks closer to windows where possible, and fit blinds if you need to cut out the sun's glare. These blinds could also help reduce the temperature in the summer by blocking out direct sunlight and help insulate the office in the winter. 'Staff provided with an outdoor view exercise their eyes more, due to the adjustment in focal length when looking outside,' says Adrian Norman, a BREEAM (Building Research Establishment Environmental Assessment Method) and LEED (Leadership in Energy and Environmental Design) accredited designer at office interior design specialists Morgan Lovell. 'Remove clutter from windows to let in light and add task lighting to each desk. If natural light is not available, skylights and sun tubes, which use mirrors to reflect light from the outside down a tube, offer a cost-effective method of providing natural light in otherwise dimly-lit areas.'

'A large car with one passenger emits more carbon per person (41.4kg) over a distance of 100 miles than travelling by train (9.3kg), coach (4.8kg), plane (27.5kg) or a smaller car (20.5kg).'

Complete your journey

According to Act on CO_2, a large car with one passenger emits more carbon per person (41.4kg) over a distance of 100 miles than travelling by train (9.3kg), coach (4.8kg), plane (27.5kg) or a smaller car (20.5kg). So, it makes sense to reduce office car use where possible. Consider setting up a car sharing scheme so that colleagues can take it in turns to drive each other to the workplace. A team of four would be able to take three cars off the road during the week this way. If colleagues are setting up meetings, ask them to consider doing it via videoconferencing and avoiding travel all together. It's quick and easy to set up, speaks volumes about your company's commitment to doing the right thing – plus they'll save the time spent travelling too. If you run a car pool and your company is prepared to make an initial investment, you could think about replacing conventional cars with hybrid varieties. The government's Enhanced Capital Allowance scheme (ECAs) enables businesses to write off the entire capital cost of their investment in certain energy efficient equipment – including cars with low CO_2 emissions.

Sobering statistics

According to Nicky Amos of Nicky Amos CSR Services Ltd:

- Small- and medium-sized businesses in the UK spend £6 billion on energy every year, of which £1 billion is wasted through inefficiency.

- The average company spends around £5,000 a year on electricity, yet £1,000 of this can be saved through energy efficiency measures.

- Leaving a window open overnight wastes enough energy to power a small car for 35 miles.

- 95% of the electricity from a mobile charger escapes when the charger is left plugged in after the phone has been removed.

- Lighting an office overnight wastes enough energy to heat water for 1,000 cups of tea.

- A compressed air leak the size of a match head wastes enough energy to toast 444 slices of bread per day.

Summing Up

If we're not at home, the chances are we're at work. Combine this with the fact that too few people act environmentally responsibly at work, and you have one big problem. Surveys have shown that it seems when we enter the office we leave our green halo at the door, but while changing settings on office equipment, getting savvy with paper wastage and using green stationery will have no bearing on the success or failure of your office – it will make every difference to our planet.

Of course, within an office there are also the legal requirements to consider and the fact that customers want to do business with ethically sound companies. Take a close look at your office and how its procedures and operations could be improved upon to ensure they have as little impact on the environment as possible.

Within an office there are so many areas that could be improved upon in terms of their environmental impact – from transport and office supplies, through to lighting and computing equipment. Once you've identified those anti-green culprits, take measures to stamp them out and let your colleagues know what you're doing so they can follow your lead.

Chapter Five

Going Green in the Kitchen

Waste not

Food waste can be dealt with at home by starting a compost heap. Not only will this recycle your food waste but it will also have a big impact on the growth in your garden. Georgina Wroe, allotment enthusiast and former editor of *Grow Your Own* magazine comments: 'From Prince Charles to the local allotment society – everyone is doing it. Making your own compost is the latest craze to sweep the UK, as gardeners catch the grow-your-own bug while embracing the recycling message.'

You can either start a heap in your garden or purchase a compost bin that will keep your garden tidy. It needs to be situated in a fairly sunny part of your garden and should be made up of around 50% 'green' waste and 50% 'brown' waste. It can take up to a year before your compost is ready and when it is it should be crumbly and dark and a little moist. 'Home composting is a vital part of a healthy garden, or allotment,' Georgina continues. 'You can recycle vast amounts of waste that would otherwise end up as landfill, and create a sustainable supply of rich, organic matter to enrich your soil. But, to some gardeners, the secrets of perfect compost are on a par with nuclear fusion. Ratios of carbon and nitrogen, plus what you can and can't compost, leave many gardeners as cold as a badly-tended heap.'

- Green waste: tea bags, grass cuttings, vegetable peelings, fruit waste, coffee grounds and filter paper, old plants.

- Brown waste: cardboard, egg boxes, twigs, branches, bark, scrunched up paper, leaves, sawdust.

Don't put any of the following on your compost heap:

- Rocks, bricks or rubble.

- Human or pet waste.

- Diseased plants.

- Cooked food (including vegetables).

- Meat.

- Dairy products.

Food figures

According to the Love Food Hate Waste campaign, every year UK households throw away 8.3 million tonnes of food. Not only is a significant proportion of your weekly shop ending up in the bin (the campaign estimates the average family bins as much as £680 worth of food annually), but there's also a serious environmental consideration too. According to the campaign, if those 8.3 million tonnes of wasted food ended up on our plates instead, we'd be able to slash carbon emissions by the equivalent of taking one in four cars off the road.

Top tips for reducing food waste

- Avoid 'buy one, get one free' offers on fresh food unless you're certain you can use it. It's only a money saving find if you end up eating it.

- Keep your shopping list in your kitchen and add to it when you use up a product. This way you're only buying what you need.

- Keep your fridge at the optimum temperature (between one and five degrees) to ensure your chilled food stays fresh for longer.

- Get creative with your leftovers – vegetables can be blitzed into a soup, stale bread can be turned into breadcrumbs, too ripe fruit can be used for smoothies, the remains of the Sunday roast could be used to make a homemade stock.

- If you've got something in the fridge that will soon be out of date and you're not sure what to do with it, get some inspiration from a recipe website.

- Once a week sit down and plan your meals for the week ahead – this will help you reduce the amount of food you buy that goes uneaten, so reducing waste and your supermarket bill.

Be a green fingered goddess

The ultimate way to cut down on food miles, packaging and energy is to grow your own. And space needn't be an issue – even a tiny window box will still yield some herbs for use in the kitchen. If you've never done it before then start small by dedicating an area of the garden to a few vegetables. You can also buy simple grow bags for items such as potatoes, which can be grown on a patio. There are many books available to help you know what to do and when, so get down to your local library and get reading up on when to plant your veggies.

'Don't think you need acres of garden, or even an allotment to grow great-tasting veg,' reassures Georgina. 'With a little imagination, you can come up with a creditable harvest from a patio. If you are daunted by growing from seed, many mail order seed companies send out plugs, or young plants. Plus it is a fantastic way to get children interested in the seasonality of food and out into the garden. Supermarket vegetables are all about shelf life and nothing about taste. Nothing is as delicious as a just-dug potato or freshly-picked tomatoes – all free from pesticides.' From your garden straight to your plate means the only energy involved in processing the food is that of you digging them up – there's zero packaging and the food is as fresh as it can possibly be.

Greening over your weekly shop

By now most people are aware of the need to use reusable bags instead of plastic carrier bags, but what else can you do to ensure your weekly shop is as environmentally friendly as possible? The first thing is to look at where you shop – is there a local butchers or greengrocers that you could cycle to? Find out if there's a fruit and veg market in your nearest town and take advantage of it. By supporting local suppliers you're also supporting the environment and reducing the food miles involved in your menus.

What are food miles?

Food miles refers to the impact our food has on the environment by calculating how far it has travelled from the point of production until it reaches the consumer. So if your apples come from South Africa, they will have notched up many miles of travel to get to you, whereas there may be an equally acceptable type of apple locally grown and therefore fewer miles involved in their transport to you.

Every year food transportation contributes nearly 19 million tonnes of CO_2 to the atmosphere, so do your bit to reduce this by opting for local and seasonal ingredients. Don't forget that by automatically leaping in your car to get to the shops you're also contributing to this burgeoning CO_2 figure. If you can, why not walk to the shops – a shopping basket can help with any heavy bags, or you could try cycling and fitting a box or basket to your bike. If the distance involved is too far to cycle or walk, think about jumping on a bus instead.

Pack it in

When you're at the shops, take a look around and pay attention to just how well packaged your food is. Certain foods need to be adequately packaged to protect your food and ensure it meets stringent hygiene and safety standards, as well as being a vehicle for displaying important information about your food. But don't be fooled – a lot of this packaging is unnecessary and adds to your

42

food bill and your waste. In order to create this packaging, natural resources such as energy and water are used, plus there's the additional issue of how long different types of packaging take to break down. In particular, plastic is difficult to recycle and can sit in landfill sites for hundreds of years.

There are several steps you can take to help limit the packaging waste you have. Buy items in concentrated forms; tomato puree, fruit cordials, washing powder – they're all available in concentrated form which means you use less and so have to replace the item less frequently. Also look out for refillable options. Some brands make available refill bottles that typically use less packaging (or recyclable packaging) and are cheaper. Get to know what's available in your local town too – there are some shops that allow you to buy according to weight and take your own bottles and bags for refilling purchases.

'Don't fall into the trap of thinking that all packaging is evil – it's not. Packaging can actually play a vital role in reducing our food wastage on a global scale.'

Fast fact

Don't fall into the trap of thinking that all packaging is evil – it's not. Packaging can actually play a vital role in reducing our food wastage on a global scale. As an example, cucumbers wrapped in plastic will remain fresh for up to two weeks – without the wrap that cucumber would end up in the bin in a matter of days. Packaging is necessary – but excessive packaging should be stamped out.

Whenever possible, buy items loose – this is particularly true for fruit and vegetables which can be bought either loose or covered in layers and layers of plastic. If it's inevitable to buy something in packaging then take the time to look at all your options. Does the brand you normally buy use lots of excess packaging and is there a similar product that uses less packaging? Is the product you want available in packaging that is easier to recycle such as cardboard? Can you buy the item in bulk (as long as you know you'll use it) to reduce how often you need to buy it?

Swap shop

Next time you hit the supermarket or nip out in your lunch break, think about the easy swaps you could make.

Swap...	For...
Supermarket sandwich in plastic casing	Bakery sandwich in paper bag
Bottled water	Reusable water bottle
Jar of coffee	Recyclable bag of coffee as a refill
Bottle of milk	A milk bag and reusable jug
Eggs in plastic containers	Loose eggs and take your old cardboard egg box

Know what you're buying

Food labels can be confusing and when it comes to the environment it's hard to know which label is important. Fair trade works by giving small-scale producers a guaranteed price for the product to ensure that no one is being exploited in the process of producing items. Organic certification is required by law for food and drink labelled as organic – artificial fertilisers are banned alongside controversial additives including aspartame, tartrazine, MSG and hydrogenated fats, and pesticides are highly restricted.

So what does organic food mean for the environment? 'Organic food is produced from natural and sustainable farming systems which avoid the use of pesticides and which prohibit the use of artificial fertilisers and GM technology,' explains Clio Turton, press office manager for the Soil Association. 'Not only is organic food rated best for animal welfare by Compassion in World Farming, it also contains higher amounts of beneficial minerals, essential amino acids and vitamins. Recent European research shows that it is mainly artificial fertilisers that depress beneficial nutrients in fruit and vegetables, so generally all organic food will contain more healthy nutrients.'

'Organic farms have 50% more wildlife, support more and better farming jobs, and causes less pollution,' she continues. 'Crucially, as the world faces the terrible threat of climate change, organic food generally has a lower carbon footprint than non-organic. Government targets state that we have to cut greenhouse gas emissions by 80% by 2050 – research shows that organic farming systems can take carbon out of the air and put it back safely in the soil, fighting global warming while at the same time producing healthy food.'

Organics on a budget

Some organic food does come at a premium, but if an organic product does cost a little more, you really are getting what you pay for – there are none of the hidden costs that are associated with intensive food production. So, with a little bit of forward planning and creativity in the kitchen, going organic does not have to add hundreds of pounds to your annual food bill. The Soil Association has some useful advice for anyone keen to go organic without increasing their weekly shopping budget too much.

They suggest signing up for an organic box scheme where local, seasonal and organic fruit and veg is delivered direct to your door. To find such a scheme, use a search engine and type in your local area and 'organic box scheme'. If you tend to stick to convenience food, consider preparing meals from scratch with organic produce. You'll save on the cost of ready meals, even though the raw ingredients may be a little more expensive in organic versions. Look at which organic ingredients are cheapest. For example, organic meat can cost considerably more than non-organic, so try to cut back on how much meat you consume and increase your fish, pulses and vegetable intake. The Soil Association also advocates joining an organic buying group. Through this you can bulk-buy store cupboard staples with a group of friends at wholesale prices. If there's not a local group to you, why not create one? The Soil Association has a useful guide to doing so on its website (see help list). If you can't afford to buy organic, support local farmers' markets and shops to keep food miles to a minimum.

Price comparison

The table below shows a rough guide to the price you can expect to pay for comparable organic and non-organic items from supermarket own brands.

Item	Non organic	Organic
Bananas	£1.34	£1.69
Whole chicken	£3.49	£5.99
Milk	£0.86	£1.03
Beef burgers x 4	£2.12	£3.00
Lettuce	£1.07	£1.38
Rice	£0.68	£1.17
Potatoes	£2.29	£2.89
Bread	£1.35	£1.49

'If you're keen to go green but need to keep a tight leash on your food expenditure, why not opt to swap a few items for organic versions?'

Top 10 organic buys

If you're keen to go green but need to keep a tight leash on your food expenditure, why not opt to swap a few items for organic versions? There are certain foods that will have the biggest positive environmental impact if you opt for organic varieties. According to the Pesticide Action Network UK (PAN UK), the top 10 foods that are more likely to be affected by pesticides are:

- Flour.
- Potatoes.
- Bread.
- Apples.
- Pears.
- Grapes.
- Strawberries.

- Green beans.
- Tomatoes.
- Cucumber.

What else can you do?

As well as buying some organic produce where necessary, you should also check the labels on fish products. The Marine Stewardship Council logo will assure you that the seafood you're buying has come from sustainable sources. This means that the seafood is caught from stocks that aren't being over-exploited as well as ensuring that the fishing methods used won't damage, harm or destroy the marine environment for other creatures.

Get cooking savvy

The mode of cooking you use can also be an energy saver. By adopting more savvy methods of producing your meals, you can save water and energy without impacting on the taste of the final product – plus it could even save you time and money! Cooking accounts for around 3% of your total home energy use – but while that may not sound like a lot, the use of various cooking appliances (cookers, kettles, microwaves, etc) uses as much electricity in a year as British street lights use in six years.

Gas ovens are typically more environmentally friendly than other types of ovens. Although gas uses more energy, it has a lower CO_2 output making it a greener choice for cooks and cutting your carbon emissions considerably. Your microwave is also an ally in greener cooking – it will use less power to cook frozen veggies compared to boiling them on the hob. They also use on average half that of the power of an electric oven – though this does depend on what you're cooking. If your microwave has a digital clock, it's worth switching it off at the plug to save energy and using a battery-powered (rechargeable of course) clock day-to-day instead.

Top tips for hob cooking

If you're using your hob, keep in mind these top tips: use just enough water to cover the contents of the pan, and always put a lid on it to trap the heat and speed up the time it takes to cook. Don't automatically grab the largest saucepan and stick it on the biggest cooking ring if you're only cooking a small portion. Use a pan suited to the size of the food you're cooking and a suitably sized cooking ring to reduce wasted energy.

Be water efficient

It's a resource that we've come to rely upon without really thinking about the impact it has on the environment. How many times have you left the kitchen tap running while you run to answer the phone or washed up by leaving the tap running continuously rather than filling a washing-up bowl? But wasting water is a real environmental no-no, so think carefully about how you use it in the home. A dripping tap could slowly but surely waste up to 20 litres of water every day, costing around four pence daily. A new washer is easy to fit, costs little and will stop you wasting this valuable resource.

'A dripping tap could slowly but surely waste up to 20 litres of water every day, costing around four pence daily.'

Simple solutions

If you're guilty of running the tap before you fill your glass, you could also be wasting up to 10 litres of water every day. If you like your water cold, fill a jug and keep it in the fridge – chilled water on 'tap' without running it to cool down first. When washing and peeling your veggies, do them in a bowl of water and then reuse this water to water your plants so every precious drop is put to good use. When using appliances that run on water, such as the washing machine or the dishwasher, always ensure you have a full load before starting it. By doing this you could save around five litres each day per appliance. Also, check to see if the appliance has an eco-setting that will further enhance your eco credentials.

Finally, get in touch with your water supplier to find out what they can do to help. Many companies have water saving packs or devices for customers which you can request free of charge. Thames Water offers its customers a free cistern device to reduce the amount of water you use when you flush the toilet, while Anglian Water offers its customers a ShowerSave which reduces the amount of water you use when showering and could save you up to £46 on your water and energy bills.

Summing Up

For many, cooking is a great source of pleasure – but have you ever really questioned the origins of your food? Greening over your kitchen is all about looking where you source your ingredients from – local is best – and whether they're approved by any eco organisations. Organic is the golden seal of approval for chemical-free produce.

Packaging is another area of concern – some supermarket produce may be triple wrapped – with a plastic tray, plastic wrap and cardboard pack. It may well help to protect our food, but all that packaging has to go somewhere – and more often than not its final destination is landfill. Get into the habit of asking questions about each purchase to assess its green credentials and ensure everything in your shopping basket is as kind to the environment as possible.

Becoming cleverer with your cooking techniques could knock pounds off your energy bills and help to remove the stain that is your carbon footprint. Think about the water you use, the pots and pans you have at your disposal and whether your microwave could do a more effective (and greener) job.

Finally – why not roll up your shirt sleeves and get dirty in the garden? If it's good enough for Prince Charles, it's good enough for us! Growing your own fruit, vegetables and herbs will dramatically cut down on harmful pesticide and fertiliser use and give you a veritable bounty of tasty treats at your disposal.

Chapter Six

Going Green When Travelling

Transport counts for nearly a fifth of the UK's total domestic CO_2 emissions – but with a few simple tweaks you could drastically cut your carbon footprint. So which forms of travel are bad for the environment? The obvious mode of transport that springs to mind is plane travel. However, carbon emissions from cars are just as much of a problem, not least because we tend to use our cars so heavily. In fact, the best way to slash that burgeoning carbon footprint is to use your own steam to get around – by walking or cycling.

Plane sailing

Our insatiable thirst for foreign climates is pushing up our carbon footprints – a return flight from London to Barcelona will emit 277kg of carbon emissions. But if you absolutely have to fly, how can you make it as eco friendly as possible? The first thing is to consider when you fly. A project from the University of Leeds found that night flights have twice the impact on the environment as day flights, due to the effect of a plane's condensation trails.

So, once you've booked your day flight, it's time to consider your route. If you're booking a short-haul flight (within Europe, for example), always try to find a route that flies direct, so there's no connections or stopovers. A plane uses up the most energy at take off and landing, so where possible avoid increasing the number of times you do this on a trip. There is an exception to the rule though. For long-haul flights, a massive amount of energy is eaten up just in carrying the fuel, so it becomes more energy efficient to include a stopover or two.

Carbon offsetting

Carbon offsetting enables you to compensate for your carbon emissions by helping to fund projects that create an equivalent CO_2 saving, and is commonly used in the travel industry. Of course, the best way to make up for your CO_2 emissions is to reduce them in the first place, but sometimes a flight or car journey is unavoidable. There are numerous carbon offsetting schemes available, but the best way to ensure that the money you donate goes right to the heart of a project is to chose one which has met the government's strict criteria of the Quality Assurance Scheme for Carbon Offsetting.

The scheme guarantees that your emissions will be accurately calculated, prices will be transparent, you'll be provided with important information related to cutting your carbon footprint and that carbon credits purchased comply with the Kyoto Protocol (an international agreement that sets targets for countries to reduce their emissions) and have been verified by either the United Nations or the EU's emission trading scheme. You can find a list of approved companies at www.actonco2.co.uk.

Most schemes have a carbon calculator you can use online to work out your total carbon footprint, or just the footprint of a particular journey. If you're using a carbon offsetting scheme that is not quality assured then always make sure that the money you donate represents a real reduction in carbon emissions and has been verified by a third party.

'With a quarter of all the car journeys in the UK under two miles, digging out your old bike could be the way forward.'

On your bike

With a quarter of all the car journeys in the UK under two miles, digging out your old bike could be the way forward. A two mile journey on two wheels would take less than 10 minutes for most people – and of course the more you cycle, the fitter you'll be, and so the quicker you'll be able to go in the long term. You certainly don't need a fancy bike – as long as it has two wheels it will do the job, but do get your cycle serviced regularly to make sure it's safe to be on the roads and always wear a cycling helmet for your own safety.

If you don't have your own bike and are based in London, the capital is home to a new cycle-hire scheme. Six thousand bikes have been distributed across 400 central London locations and will be available 24-hours a day, seven days a week, with no need to book. It's worth contacting your local authority to see if they have any similar schemes in place or searching for a bike hire facility in your area.

Ditch the car

Lose the car and you could also lose a significant chunk of your carbon footprint – but many people still need access to a vehicle. So what options are there for part-time car users?

Car pools

If you know you still need to be able to have access to your car then think about how you could use it in a more energy efficient manner. If you use it for the school run or to get into work, find out who else is local to you and could share the journey – it sounds obvious but a full car will generate less than a quarter of the emissions of five individual cars all doing the same journey. Car pooling is a popular option and there is a host of websites springing up to help match you to your perfect carpooling partner. Liftshare, National Car Share and Carplus can all help you find people doing a similar journey as yourself (see help list).

Car clubs

Car club schemes are also becoming more popular. Zipcar, a car club scheme operator, estimates that each of its Zipcars takes up to 20 personally owned vehicles off the road. So how do they work? Carplus, a charity promoting the benefits of car clubs, estimates that there are around 28,000 car club members in the UK who are sharing approximately 1,000 cars. The system works through a membership programme – some schemes charge an annual fee or a joining fee – and once you're a member you can use any one of the available cars paying an hourly rate. The cars are kept in designated bays so you know where to find them and you can use your membership card to access them.

The cost is very attractive to members – with Streetcar, one of the largest car clubs, you pay £49.50 a year membership and can then hire a VW Polo midweek for just £3.95 an hour, or £82 for 24 hours. These charges cover the cost of insurance, breakdown recovery service, maintenance costs and road tax, but you're still liable to pay for the petrol you use (with Streetcar this is 19p per mile but you get 30 miles' worth of free petrol each day, so a short journey could cost you nothing in petrol). If you're happy to walk, cycle or use public transport and only use a car club when you really need it, this scheme can work well.

Think before you drive

If you do need to drive then a few amendments to the way you do so could see you cutting your fuel use by as much as 4%. Fleet Department, a vehicle management company, offers the following advice on achieving this:

- Aggressive braking increases toxic emissions by more than five times and fuel consumption by as much as 4%. The best way to drive is to accelerate slowly and smoothly, then get into a high gear as quickly as possible.

- Increasing your motorway cruising speed from 55mph to 75mph can raise fuel consumption as much as 2%. You can improve your fuel mileage up to 15% by driving at 55mph rather than 65mph.

- Under-inflated tyres can cause fuel consumption to increase by as much as 6%. Check tyre pressure at least once a month, when the tyres are 'cold'. Most petrol stations have public-use air compressors with a pressure gauge, so you can make a point of checking when you fill up.

- Remove any unnecessary items from inside the vehicle when not in use to decrease the weight. An extra 100lbs (48kg) of weight can increase your fuel bill by 2%.

- Dirty air filters can cause your engine to run at less than peak efficiency. Regular visual checks of the air filter will tell you if it needs replacing. Clogged filters can cause up to a 10% increase in fuel consumption.

Public transport

Public transport has really suffered with the prevalence of cheap, second-hand motors, but using alternative means of transportation can slash your carbon footprint and save you money. It also reduces the hassle and frustration of trying to navigate around an area you don't know, fighting for a car parking space, or forking out a fortune for petrol, car tax and the glut of repairs your car will need.

Using buses or trains helps to reduce the volume of traffic on the road, which will also have a positive impact on congestion (think how many cars you see with just one person in them fighting through rush-hour traffic). Public transport also provides a far more effective fuel efficient form of travel for passengers, compared to that offered by car journeys. However, delays, geographical restrictions and frequency can mean public transport is not always the best option.

Trains get greener

The UK government has started work on a £1.1 billion rail electrification scheme of the Great Western mainline, which should cut carbon emissions considerably when it's finished, expected to be 2017. Once complete, trains will be replaced with either Super Express electric trains or a hybrid/electric train – electric trains emit 20-35% less carbon per passenger per mile compared to conventional trains. Currently, only around 33% of the UK's rail network is electrified, but the government plans to increase the use of electric trains to around 60-65% in the future.

On the buses

It's not only buses that are having a green facelift – bus networks are also becoming ever greener. In London a new generation of buses are hitting the streets which promise to be greener than their conventional counterparts. The hybrid buses are ultra-quiet, low carbon vehicles and by 2011 are expected to represent just under 4% of the London fleet, with 300 in total. Featuring a diesel-electric motor, they'll cut CO_2 emissions by as much as 40%. If you

'Public transport also provides a far more effective fuel efficient form of travel for passengers, compared to that offered by car journeys.'

fancy giving them a try, head for route 141 London Bridge to Palmers Green which features a Volvo I-SAM (Integrated Starter, Alternative Motor) system, similar to the Toyota Prius.

Stagecoach Group plc is also working hard to improve the eco credentials of its fleet by investing £70 million in greener buses for the UK network. Each of the new buses meet the new Euro 5 emissions standards which came into force in September 2009 and have been adopted by the EU to cap pollutant emissions.

Be a responsible traveller

Just because you've made a commitment to going green does it mean you have to forgo your two weeks in the sun. You can still enjoy a holiday (and why not, you've earned it), but to ensure that your break doesn't do more harm than good, follow the six golden rules of eco travel:

Holiday close to home

Your annual break doesn't have to see you heading to far flung destinations – holidaying close to home is one way of reducing your carbon footprint. Justin Francis, co-founder of responsibletravel.com says: 'Choosing to holiday at home, assuming we don't take a domestic flight, creates significant environmental, social and economic benefits. We all get to appreciate local distinctiveness on our doorstep and at the same time reduce our overall carbon emissions. Holidaymakers will increasingly discover that the exoticness of the unknown doesn't have to take the form of a desert island in the middle of the Pacific.' A bit of digging around and you can find some truly unforgettable breaks including surfing in Cornwall, a tree climbing tour of the Isle of Wight or sea kayaking in Wales.

Try a different mode of transport

Think holidays and most people think about jetting off, but who wants to start their holiday stuck in the departure lounge at Heathrow? Europe has an extensive rail network and train journeys have a considerably lower impact on the environment than flying or driving. A domestic flight will spew out 180g of carbon per traveller per kilometre, while rail travel emits just 5.7g, beating car travel by 117.7g.

Be aware of greenwashing

With 'eco' being the new buzzword, don't be surprised to find that, amongst the very credible outfits, there are also those that are looking to exploit their marketing potential for financial gain. Consumers are actively searching out eco-friendly alternatives, and so some untrustworthy hotels may spend thousands on ad campaigns trying to convince consumers of their ethical practices when in reality it may amount to little more than a recycling bin and the option for travellers to reuse their towels. A survey by Deloitte in 2008 revealed that almost a quarter of travellers are prepared to pay extra for a green hotel, and, unsurprisingly, many hotels – green or not – want to help you part with your cash. See chapter 1 for more information on greenwashing.

Support the local economy

If you're going to be heading for the deepest, darkest depths of Asia then make the effort to experience the area as a local. Treat the locals with respect and ask them where they eat and shop and head there – this way you can be sure your money is going direct to the local economy. Avoid buying your favourite brand of coffee/cornflakes/chocolate as this will have most likely been imported involving huge air miles. Most package holidays have a range of tours available for tourists but, if you can find one, a local guide will be much more knowledgeable about the area, its culture and attractions, plus you'll help a local earn an income.

Give something back

Consider something different from your usual beach holiday and opt for a volunteering holiday. The opportunities really are endless – from elephant conservation in Thailand and teaching English to children in Ghana through to coral reef conservation in Honduras or rebuilding walls in England. You'll also get an authentic taste of what life is like in your host country by experiencing the culture, the locals and the problems first hand. But beware of what proportion of your costs actually go direct to the local community. 'Voluntourism' can be big business and while your hard work does directly impact positively on the local environment, you may find the hefty fee you stump up to join the programme lines someone's already-wealthy pockets.

Protect valuable resources

'When abroad, use water as sparingly as possible, eliminate lengthy showers, don't leave the tap running and don't hand wash items individually.'

In England the volume of rain means we rarely experience a serious drought, but in many countries water is a valuable commodity. When abroad, use water as sparingly as possible, eliminate lengthy showers, don't leave the tap running and don't hand wash items individually. In fact, it is good practice to adopt these habits at home to help reduce future impact of droughts. You should also avoid contaminating the local water supply by ensuring you're using chemical-free products such as shampoo, shower gel and conditioner. Likewise, also be considerate through your use of electricity – switch off lights and appliances when you leave and always pack a solar-powered multi-purpose charger for phones, palmtop computers and iPods. And if you don't change your towels and bedding everyday at home, do you really need to have them changed daily on holiday? When you arrive, advise the reception desk at a hotel that you're happy to have them changed twice weekly or weekly instead.

Summing Up

Transport and our heavy use of it means that it could add up to nearly 20% of your carbon footprint – so becoming an eco traveller would really help to minimise this. Often the act of jumping in the car is done without thought – but always ask yourself whether the journey could be made using a different form of transport?

The good news is going green doesn't have to mean having to sacrifice your car – it just means getting smarter with how you use it. Try cutting back how often you use your vehicle – try walking, cycling or even rollerblading for some of your journeys. Because of the UK's congestion, you might even find you make it to your destination quicker.

Holidays can also be given a green makeover to ensure you don't have to give up your good intentions for two weeks of the year. Being a responsible traveller goes hand in hand with eco issues – so protect the local environment, respect and support the locals and think about swapping some of your foreign jaunts for exploring parts of beautiful Britain.

Chapter Seven

Green Beauty

Be a natural beauty

Going green in your skincare routine is easy – if you know what to look out for, what to avoid and why it's important. In pursuit of beauty, many of us are risking our health everyday – oblivious to the chemicals that are used in our skincare products. The average woman's everyday routine of showering, hair washing, styling, skincare and make-up contaminates the body with a staggering 515 chemical compounds every day.

'In the cosmetic jungle out there, a couple of "beasts" I would avoid are propylparaben and butylparaben (these are the most common preservatives used in cosmetics, especially make-up),' says Dr Barbara Olioso, a green eco chemist. 'Why? Because of tests showing hormone disrupting properties in 2007, the JECFA (The Joint Food and Agriculture Organisation and World Health Organization Expert Committee on Food Additives) banned them from food and for the same reasons the European scientific committee (SCCS) is reviewing their safety to question their ban in cosmetics too.'

How safe are your products?

Of course, there are a number of guidelines and laws in place to protect consumers from harmful chemicals, but do these laws go far enough? Many green enthusiasts would say no. By law, cosmetic companies must list all of their ingredients, but as the pressure is on to maintain sales, companies are wising up to our love of all things green and trying new tactics. Green beauty virgins need to read up on chemical names to ensure they can spot the baddies. For example, tocopherol sounds like a chemical but generally refers to vitamin E, making it harder to decipher what's really in your favourite beauty treat. As a rule, look for products that don't have extensive ingredients lists – the fewer ingredients used, the better.

Ask questions

Even once you've waded your way through an ingredients list, there's still a risk that the product you're holding is not as pure as it may lead you to believe. Manufacturers are not obliged to reveal impurities or chemicals that are present in the raw materials or used during the manufacturing process provided they don't appear in the final product. You also need to see past the packaging. If something claims to be 80% organic then you need to question the non-organic 20% and the sheer volume of harmful chemicals that could still find their way into your bloodstream.

'If something claims to be 80% organic then you need to question the non-organic 20% and the sheer volume of harmful chemicals that could still find their way into your bloodstream.'

Going organic

Going organic isn't just for trendy A-listers, there are plenty of organic brands that won't break the bank and plenty of reasons why you should opt for organic.

Label fables

Organic beauty buys are generally certified by the Soil Association in the UK (the largest, though not the only, approved certification board), so where you see its seal of approval, you can be sure that your product contains no petro-chemicals, GM ingredients, parabens, aluminium, cosmetic solvents, sodium lauryl sulphate or zirconium, amongst other chemical nasties.

Wise up

If a product contains the Soil Association logo, you can be certain that it contains a minimum of 95% organic ingredients. If a product uses at least 70% organic ingredients but less than 95%, it can still carry the Soil Association's logo but must also state 'Made with XX% organic ingredients' so the consumer can clearly see what they are buying. Additionally, the remaining ingredients must be proven to be non-GM and may only be used in beauty products if an organic version of that ingredient is not yet available, or if it is from a restricted list of synthetic products which have been assessed to ensure they do not have a detrimental effect on human health and have a minimal environment impact.

Is organic always better?

The popularity of organic and natural products has skyrocketed over recent years – but this has caused experts to express concerns for the impact it may have on the environment. Palm oil is top of the list of ingredients that cause environmental damage and as demand grows for this ingredient in organic and natural products, more rainforests are being destroyed to meet production. According to Dr Olioso: 'A cosmetic and food ingredient that is linked to big environmental damage is palm oil, causing huge deforestation in South East Asia, soil erosion and wild habitat destruction at the expense of orang-utans. The trend towards deforestation is increasing dramatically every year so I encourage you to read the ingredients list of your food and cosmetics, looking for palm oil, before you buy.' But while you may be committed to looking out for potentially catastrophic environmental ingredients, it's not always a straightforward case of reading the labels. Dr Olioso says: 'Spotting palm oil derivatives is tricky; however, if you are really committed to make a difference, ask the question "Does it contain palm oil derivatives and roughly how much?". At present it is challenging to avoid these ingredients in cosmetics, so you need to understand that and at the same time show your concern and choose products with this awareness.'

So when selecting products, ensure that not only are they natural but the ingredients are sourced from sustainable sources. It's also worth looking out for fair trade logos so you can be sure that the farmers who work with the raw ingredients, and everyone involved in the production of the beauty product, is fairly paid and treated well.

'Palm oil is top of the list of ingredients that cause environmental damage and as demand grows for this ingredient in organic and natural products, more rainforests are being destroyed to meet production.'

Know your chemicals

Bad press has meant that more and more women are wising up to certain chemicals and giving them a wide berth. Not so long ago parabens were portrayed in the press as being cancer-causing (parabens are oestrogen mimics and may increase the risk of breast cancer), but this is just the tip of the iceberg. Check your labels and also look out for:

- Sodium lauryl sulphate – this is what gives products a foaming action but worryingly it's used commercially in engine degreasers – an indication of how harsh it is. What's more, it's also been linked with damage to the eyes, brain, heart and liver, and scientists warn that it is rapidly absorbed and retained.

- Phthalates – two phthalates have already been banned in the EU because of the risk they pose to developing reproductive organs and so they are of particular concern to pregnant women. They are used in perfumes and hairsprays to slow down the rate at which they evaporate.

- Liquidum paraffinum/petrolatum – this ingredient is a by-product of petroleum and is used widely because it's cheap. It works by coating the skin, which in turn can clog your pores. It interferes with the skin's ability to eliminate toxins and slows down the skin's function, which may result in premature ageing. It's often used in lip balms and body oils.

- Formaldehyde/methyl aldehyde – yes it's used to preserve bodies, but it's also a common ingredient in hand washes, deodorants and nail varnishes. It's used in the beauty industry as a disinfectant and a preservative but it's suspected of being cancer-causing and in sensitive individuals can trigger an asthma attack or skin irritation.

- Para-phenylenediamine (PPD) – if you regularly dye your hair, check the box carefully. PPD is still used in some dark hair dyes despite the fact that is has been linked to a higher incidence of bladder cancer, scalp conditions and skin allergies.

How will my skin respond to natural products?

Becoming a natural beauty doesn't mean you need to compromise on the quality of your products or the results you can expect to see. There are many high quality products out there that can deliver beautiful skin without the need for harsh chemicals. Making the move from synthetic beauty products to natural versions is easy but it may take some time for your skin to adjust. Your skin may need to learn to regulate itself instead of relying on chemical products to do so, and during this time you may experience breakouts and blemishes. However, within a week or two your skin should be glowing with health in ways you've never experienced before.

Reduce your body burden

So now you know the damage that you may unconsciously be doing to your health, what can you do about it? Well apart from giving up washing and refusing all beauty products, there are small changes you can make for big results.

- Take a close look at how many products you use and decide if they are all strictly necessary. Is your bathroom cluttered with toners when a simple splash of water will do?

- Go back to basics – avoid products like baby wipes, which may contain parabens and propylene glycol – a common ingredient in anti-freeze – and instead use damp organic cotton wool.

- Read the labels of everything you buy and always try to find the most organic, natural product you can.

- Many women are guilty of over-washing – twice daily showers are rarely necessary – while many of us don't need to wash our hair every single day.

Eight natural beauty companies to try

The Weleda range is totally natural and totally affordable which is why it's so popular.

Green People was founded out of the desire to provide a range of products that don't irritate skin and so the first thing to be dumped was the chemicals.

New kid on the block is Nothing Nasty, with products that are wonderfully pure. The range is formulated using only natural ingredients: vegetable, nut and seed oils, essential oils, natural waxes, sea salts and herbs.

There are few celebrities who haven't proclaimed themselves to be a fan of Burt's Bees. Created using natural herbs and essential oils, there are no preservatives, chemicals or artificial colours in the range, so it's good for you and the environment.

Neal's Yard Remedies was founded back in 1981, making it the first company in the UK to sell products that were certified organic. Today it is still a popular choice with natural beauties and has a massive range of products on offer.

Your make-up bag could also benefit from a green makeover. Take a look at BareEscentuals Mineral make-up, it's so pure you can even sleep in it! The range is made of 100% micronised minerals with nothing else added.

Origins has long been a cult brand with beauty aficionados and now it has launched its own organic range. The Origins Organic range includes products for face, hair and body.

Natural Organic Edible Cosmetics are a favourite of Gwyneth Paltrow. This range of 100% natural, 100% edible products means you don't have to worry about damage to the environment.

Top tip

Remember that all of those products you use end up being washed down the drain at some point – and if they're packed full of chemicals, you're helping to contaminate our clean water supplies.

You are what you eat

Instead of stocking up on products, you could be a real natural beauty and ditch all those products and hit the greengrocers instead. Try these foods for a beauty boost:

For skin

- Kiwi fruits – kiwi fruits are a fantastic source of vitamin C. One of the key functions of vitamin C is its ability to strengthen capillary walls. By keeping the capillary walls healthy, essential nutrients and plenty of oxygen can reach the skin, which in turn optimises cell regeneration and maintains levels of elastin and collagen.

- Eggs – eggs are a nutritional powerhouse, doing everything from preventing dermatitis, reducing acne, avoiding white spots on the nails and strengthening the hair. Eggs are rich in zinc, which can help to calm problem skin by reducing inflammation and boosting skin healing. They also contain biotin, which can help to protect against free radicals and the effects of pollution.

For hair

- Almonds – it's thought that almonds' blend of nutrients can help prevent premature ageing of the hair. These nuts are also extremely rich in vitamin E, which helps to improve the elasticity of the hair and hydrate your crowning glory. It's no wonder that so many hair products on the high street contain almond oil.

- Milk – we all know milk's stellar effect on the nails, but it can beautify your hair too. Milk (and other dairy products) is rich in B vitamins, which can help to combat the damaging effects of stress. A stressful lifestyle can result in dry, lank hair, which is never a good look but can easily be turned around.

For nails

- Tuna – tuna is a rich source of vitamin D, which helps to strengthen the nail bed and promote healthy growth. It's also a great way to increase your levels of vitamin B12, which helps to boost circulation. This vitamin is responsible for giving your nails a healthy, rosy glow as it encourages a constant flow of blood to the extremities.

- Prunes – prunes are a great source of iron and people who lack this essential nutrient often complain of sallow, dull skin, lifeless hair and dry, flaky nails. Plenty of iron is especially important for vegetarians, as meat is another rich source.

'Eggs are a nutritional powerhouse, doing everything from preventing dermatitis, reducing acne, avoiding white spots on the nails and strengthening the hair.'

Quick beauty-boosting tips

- Need to get a break-out under control quickly? Dab a little fresh lemon juice on spots to take advantage of lemon's astringent properties and to remove bacteria.

- Pop some oats in a pair of old tights and hang them over the hot tap when you're running your bath. You'll be left with silky soft skin.

- Want to whiten and brighten your teeth naturally? Strawberry juice can remove stains caused by cigarettes and red wine. Just rub a halved strawberry over your teeth.

- If you've run out of body moisturiser, grab the olive oil from the kitchen and smother it over your body. For an exfoliating treat, add a handful of sugar and scrub up.

- Want tighter, smoother, plumper skin? You can't go wrong with this quick egg fix. Smooth egg white over your skin and allow to dry. Wash off and your skin will thank you for it – it's a shame it's only temporary!

- For a final hair rinse, use cold tea, beer or vinegar to boost shine and increase vitality.

Summing Up

No one wants to forgo looking good, but thanks to the prevalence of natural beauty brands you won't need to. Every day we use a huge number of beauty products – from shampoos and shower gels through to make-up and hand cream, and synthetic versions are loaded with chemicals. Switching to natural alternatives will instantly reduce the number of chemicals we're coming into contact with every day.

You can also consider how you're using these products – many of us are guilty of using too much product or using them too frequently, hoping for better results. Stick to the guidance advice on the packet so you prevent wasting product.

And if you really want to go all natural, why not ditch your favourite products and instead poke around in the store cupboard to see what ingredients are lurking that could be used in your beauty regime? Oil, lemon, eggs and vinegar are found in most kitchens and can work wonders for your skin and hair.

Chapter Eight

Green Parenting

If there's one thing guaranteed to get you switching to environmentally friendly products, it's the wellbeing of your family. But it's not just the health of your children that should move you to make changes – the environmental impact of having children is vast. Nappies, clothing, bathcare products, toys – as children outgrow all of these rapidly, it's easy to see why having children can leave a rather unsightly stain on your eco-friendly intentions.

Why are children more at risk of contaminants?

Experts believe that children are more at risk of chemical exposure than adults, a fact that has been put down to several different elements. Firstly, compared to adults, babies and young children breathe, drink and eat more in relation to their body weight. This means that they absorb more toxins in relation to their size. Children's bodies are also less well equipped to deal with chemicals as they are still developing and so the contaminants they may be in contact with can also put an extra strain on their young systems. A baby's skin is around five times thinner than that of an adult, which makes it significantly more porous and likely to absorb chemicals. If you're a parent then you'll already know the habit many babies make of putting items in their mouth – this also makes them more likely to ingest harmful chemicals.

Get real

At the moment a staggering eight million disposable nappies are thrown away every day in the UK – that's one massive landfill we're leaving behind for the next generation. When it comes to green parenting, opting for real nappies is the number one action we should all take. 'Currently in the UK three billion

'When it comes to green parenting, opting for real nappies is the number one action we should all take.'

disposable nappies are thrown away every year equating to 700,000 tonnes of nappies!' says Jon Rolls, CEO of Go Real. 'By using real nappies instead of disposables, a household can halve its weekly rubbish and also reduce their global warming impact by up to 4%.'

Disposable nappies contain a cocktail of super-absorbent chemicals to keep baby dry, a mix of glues, dyes, perfumes and bleaches, as well as paper pulp and plastics. Paper pulp is the largest single ingredient of the disposable nappy and unsurprisingly it's loaded with chemicals. The forests that are used to produce the pulp tend to be intensively managed with pesticides and fertilisers. Additionally, more chemicals are used in the process of converting the wood into pulp.

These nappies also contain plastics, such as polypropylene, which are typically produced from non-renewable crude oil resources. While some of the plastic used is biodegradable, it does require the right conditions to be able to do so – which aren't always present in British landfills. And in a bid to claim to be the driest option for your offspring, more and more nappy manufacturers are turning to sodium polyacrylate – an absorbent gel, which was removed from tampons in 1985 due to a link with toxic shock syndrome.

Money saving benefits

Switching to real nappies will not only reduce the pressure on landfill but it's also a sure fire way to save money. The advice from Go Real is: 'you can kit your baby out with all the nappies they need from the high street from just £90 (the cheapest real nappy option), add approximately £1 per week to wash them and you can still save up to £500 compared to disposable nappies.' And, if you have more than one child, you'll be able to reuse your real nappies, nappy covers, sanitising bucket and nappy bin – saving you even more money and resources!

Fast fact

According to Go Real, the latest independent research from the EA LCA report (2008) found that real nappies which are laundered at home could have a lower environmental footprint than disposables by up to 40% – but the saving does depend on how parents use their nappies. Go Real recommends for maximum environmental savings parents need to wash full loads at a maximum of 60 degrees, use A-rated machines, avoid tumble drying and use their nappies on a second child.

The pros and cons

Disposable nappies have only been available for around 60 years, so as yet it's hard to say exactly how long they will take to biodegrade. However, experts say it could be around 400 years – so it's almost certainly a legacy you'll be leaving behind for generations yet to come. But what seems to put the majority of parents off using real nappies is the perception that they're fiddly to deal with. This is in fact a myth, according to Jess Hyde, founder of Naturebotts.co.uk. 'Once you are organised it really isn't fiddly,' she explains. 'Putting the nappy on itself is hardly different to a disposable – there is the nappy and then a wrap. They're a bit stretchy as well so the fit is really good and the booster pad (if using) can be quickly folded to be in the right way depending on whether you have a boy or girl. You may use a paper liner as well, but even this doesn't get folded, just placed in position ready and waiting.' Of course there are some differences (besides the environmental concerns) with real nappies – both good and bad: 'I would probably use about three disposables in a day, but you'll use about five washables so there is a bit more changing,' acknowledges Jess. 'However, your child is likely to potty train earlier if in washables.'

Why not try… eco disposables

Eco-friendly disposables offer the convenience of a disposable nappy but with some green benefits. They're more likely to be made with eco friendly materials that are sourced from sustainable sources, plus they're unlikely to be treated with chemicals or use lots of unnecessary packaging. As an example, a Moltex Oko disposable uses 100% biodegradable materials such as unbleached wood pulp and a plant-based material for the waterproofing element of the nappy. Eco disposables will break down quicker in a landfill, but they do still emit methane – a greenhouse gas. You can also put some eco disposables on your compost heap – wet nappies that are fully biodegradable can be reused in the garden. But those that contain poo are not safe for the compost heap.

Bath time for babies

Even though they may be aimed at infants, bath time products contain vast numbers of chemicals that could be doing more harm than good to both the child and the environment. According to a briefing on toiletries for babies from WEN: 'Until they are six months old, infants lack a blood-brain barrier to prevent blood-borne toxins entering the brain: low level exposures that would have little or no effect on an adult brain can sabotage a foetal one.'

There are currently around 70,000 chemicals in existence which are used commercially. Every year sees roughly a further 1,000 added, and alarmingly many of these are termed 'bioaccumulative' – that is they build up in the body. No one can be sure of the effect these chemicals have on our children and while they have been deemed safe, the fact that many children have detectable levels of at least 300 different chemicals in their body is cause for concern.

Phthalates is one of the ingredients that is often found in baby toiletries – but don't expect to spot it on the ingredients list as it is normally contained in parfum, which by law doesn't require its ingredients to be itemised on the bottle. This chemical has been linked to early puberty in girls and has now been banned from some baby teething toys, suggesting that the experts are more cautious about its use. Products for babies should avoid parabens, phenoxyethanol (a synthetic preservative) and sodium lauryl sulphate (a foaming agent), and try to eliminate propylene glycol, polyethylene glycol (PEG), isopropyl myristate and formaldehyde-releasing preservatives such as imidazolindinyl urea.

Rethink baby toys

When shopping for children's toys, it's important to not only ensure they are age appropriate but also that their production has placed as little strain on the environment as possible. Wooden toys are excellent choices if they have been sourced from sustainably managed forests. This means that the forests used for producing wood are managed in a way that protects their future – through the practice of reforestation and management, growing and harvesting in an ethical manner.

Going green with kids

Getting children involved in environmental issues from a young age is one way of ensuring the next generation is ready to do their bit to help our planet. Here are some tips to make going green fun for all ages:

- Spend time as a family by taking your bicycles out instead of piling into the car.

- Spend one afternoon a week in the garden – planting seeds, weeding and watering can all be done by youngsters with some guidance. You could even divide the garden up so everyone has their own little space.

- Sorting out the rubbish into piles for recycling or general waste can be turned into a game for children.

- From a young age, ensure your kids understand the importance of switching off lights when they leave a room and not leaving appliances on standby.

- How about keeping your own chickens? You'll have fresh eggs on tap and the children can get involved in helping to care for the birds and collect the eggs.

'Products for babies should avoid parabens, phenoxyethanol (a synthetic preservative) and sodium lauryl sulphate (a foaming agent), and try to eliminate propylene glycol, polyethylene glycol (PEG), isopropyl myristate and formaldehyde-releasing preservatives such as imidazolindinyl urea.'

- Looking for inspiration for a rainy day? Get your recycling bin out and see what wonderful ideas the kids can come up with for new uses for your waste. Plastic bottles could be transformed into bird feeders, or empty boxes decorated for storing jewellery.

- Take the children on a nature walk and explain to them the importance of conserving our environment. Many larger parks have conservation officers, information sites or talks led by park rangers.

- Encourage children to sort through their toys and pick out the ones they no longer want and are happy to donate to charity to help less privileged children.

- Have a look at the Soil Association's website section 'Take Action' where it highlights lots of ways to get involved with environmental issues. You could organise a family trip to an organic farm, learn more about the disappearing honeybees or find local community projects to support.

- Think about setting the kids a green challenge – encourage them to make one change a week (and stick with it as they make a new change) and reward them with a day out.

Summing Up

There's something about watching the next generation grow up that makes you more determined to preserve our planet for them to enjoy. Many parents will look for the greenest options available for their children to ensure they are also the healthiest option. Organic food, toys from sustainably produced wood and chemical-free bath products are all readily available.

When it comes to parenting, perhaps one of the biggest differences you could make is the switch to real nappies. Yet many parents presume them to be fiddly and instead opt for disposables. Eco disposables are available if you want convenience and eco power, or try real nappies and see how you get on with them.

Going green is also an excellent opportunity to get children involved in environmental issues – from recycling and non-polluting travel through to growing your own veg and volunteering support for green initiatives. In just a few months, turning off lights and weeding their vegetable patch will be second nature and you'll have succeeded in shaping responsible citizens for years to come.

Chapter Nine

Quick Tips To Try Right Now

Home

- Put peat-free compost around your plants in the garden to help retain moisture and make the water they do receive work harder.

- Don't rush out to buy a brand new flatscreen TV. A 32 inch plasma TV will use up 255W, compared to 160W for the same size TV in the older cathode ray tube style. But LCD TVs are greener than a plasma screen.

- Cut the cost of gaming by ensuring games consoles are switched off when not in use. An Xbox uses up 165W, while a PlayStation 3 guzzles as much as 380W!

- Opt for rechargeable batteries in clocks, remotes and games around the home.

- Place a cistern displacement device into your toilet cistern, which could save you around two and half litres of water per flush.

- Instead of taking all of your unwanted items to the dump, think about how they could find a new home. Charity shops are always crying out for items, while websites such as Freecycle and Good News for Polar Bears can help you recycle items (see help list).

- Fit a low-flow showerhead to your shower. This will limit the volume of water you get from the fixture and so decrease how much water you use every shower and the amount you use on heating that water too.

- Use energy saving devices that will put equipment into sleep mode if you don't switch it off.

'Cut the cost of gaming by ensuring games consoles are switched off when not in use. An Xbox uses up 165W, while a PlayStation 3 guzzles as much as 380W!'

Work

- Switch your search engine to www.blackle.co.uk. Because it uses a black screen, it uses less power than a similar white-screen search engine.

- Give the office fridge a quick MOT. Pull it out and remove any dust which has collected on the coils at the back, check the door seals are still doing their job effectively and ensure that the fridge is positioned away from direct sunlight or anything generating heat such as radiators or ovens.

- Put an end to junk mail for good. Before dumping piles of unsolicited mail in the bin, call the company responsible for sending it and ask them to remove you from their mailing list.

- Convince your managing director to remove the plastic cups by the water cooler and instead ask everyone to bring in their own mugs. If the average worker uses five plastic cups a day, that's 100 cups in the average month, multiplied by the number of people in your office. And they'll all end up in landfill.

- Makeover your office boardroom. Get rid of the traditional paper flipchart and replace it with a wipeable whiteboard instead. Equip the room with plenty of refillable pens instead of biros and ensure there's plenty of scrap paper to make notes on.

- Avoid purchasing newspapers or magazines for your firm and instead get an online subscription and ask colleagues to read their favourite publications online.

- Arrange for bills, statements and invoices to be delivered and received electronically, using an online service such as www.nomorepost.com.

- Look out for recycling schemes for ink and toner cartridges – this will save you putting them in the bin. Refilling services are also available in some areas, which will also save you money.

Kitchen

- Drink tap water rather than buying endless bottles of water. UK mains drinking water is very high quality and uses roughly 300 times less energy than is used to create bottled water – plus there's no waste. If you're not keen on the tap, buy a water filter.

- Think about subscribing to an organic vegetable box. Fresh, seasonal veggies will be delivered to your door and they're guaranteed to be free from harmful pesticides and fertilisers.

- Use your kettle to boil water for cooking veggies on the hob or, even better, cook food in a microwave where possible.

- Get batch cooking. Cook up enough food to freeze and reduce the number of days you need to use your oven. When freezing food, always allow it to cool to room temperature before popping it in the fridge or freezer to reduce the power needed to cool it down.

- Your oven will effectively retain heat, so switch it off just before you're ready to serve to shave a few minutes' worth of power every time you use it.

- Keep an eye out for what your local supermarket is stocking. Most of the major supermarkets now stock an organics range and Tesco has even started to put carbon footprint labels on some of its groceries.

Travel

- Try to plan your route if you're going out in the car. Can you combine a trip so you only need to go out once (e.g. do the weekly shop before you pick the kids up from school)?

- If you get caught up in a traffic jam or pull over while someone in the car runs in to pick something up, it's always worth switching off the engine. If the engine is idling, you're still wasting fuel and emitting CO_2. As a general rule, if you'll not be moving for around three minutes, switch off the engine.

- If you've got air conditioning in your car, use it wisely. Although it's efficient at keeping you cool, it increases fuel consumption – so for town driving

'If you get caught up in a traffic jam or pull over while someone in the car runs in to pick something up, it's always worth switching off the engine. If the engine is idling, you're still wasting fuel and emitting CO_2. As a general rule, if you'll not be moving for around three minutes, switch off the engine.'

you're best winding the window down. However, if you're travelling at over 50mph, open windows will increase drag – making it more energy efficient to use your air conditioning.

- Why not join a local cycling or jogging group? Having the support of other people will help you to improve your fitness levels and it won't be long before you feel able to ditch the car for journeys under a couple of miles.

Beauty

- Try making your own beauty treatments. Look for fruits that are rich in natural oils, such as avocados which make great hair masks, or ingredients with tightening properties like eggs for facemasks.

- Instead of soaking in the bath every night, switch to showers with just one bath a week to really relax and unwind. This will save litres of precious water and the energy needed to heat it.

- Ditch your toner and just use a splash of water instead to freshen skin. Or make your own toner with a few drops of rose essential oil in a spray bottle of water.

- Forget expensive (and chemical laden) skincare creams and instead nip to your local health shop and get some Evening Primrose capsules. Break one open and use this to nourish your skin.

- Look out for products that come in glass jars as these can be recycled. Plastic pots are harder to recycle, so why not ask at your local beauty shop whether they offer a refilling service?

- Learn to read the labels of your favourite products so you know what's really going in to them. Can you find an alternative that is free from chemicals and has care for the environment at its heart?

Parenting

- If you're using real nappies then always soak them before washing using essential oils or a chemical-free solution. Both lavender and chamomile Roman are highly recommended for babies – but neither should be applied neat to babies' skin (if you do notice any reaction to the oils, stop using them immediately). This will enable you to wash them at a lower temperature (40 degrees) and still get them clean. Keep a sanitising bucket to hand so it's a quick and simple job to soak them before you load the washing machine.

- Whenever possible, always line-dry your real nappies. Not only will you reduce your carbon footprint and your electricity bill by forgoing the tumble dryer, but drying them outside in the sun will also have a gentle bleaching action.

- Look for real nappies that are made from organic cotton or hemp. Both of these materials will have been grown without fertilisers or pesticides.

- Avoid using synthetic baby care products. Either look out for all-natural versions or go back to basics. Warm water and cotton wool is better for immature skin than baby wipes, for example.

- Make your own baby food using good quality, organic ingredients. This way you can be sure what goes into it and avoid chemical pesticides and fertilisers.

- Plan fun activities around green issues to get your children involved in the protection of our planet. On bank holidays look out for local activities in the park or at farms which often centre around environmental issues.

Chapter Ten

Buying Green

Throughout this book we've looked at working with the resources you already have. While many people may be passionate about saving the environment, ultimately their purchasing decisions are dictated by their budgets. But, there will come a time when you'll be looking to purchase a new car, computer or fridge, so when that time comes, you should aim to make your chosen purchase as green as possible. Of course, the mantra to remember is only buy something if you really need to – the ultimate way to go green.

If, however, you absolutely have to make a purchase, labelling schemes can be helpful in this aspect – helping you to identify the greenest options for everything from cars and dishwashers through to bananas and flowers. Common labels used in the UK include the Fairtrade mark, the Marine Stewardship Council and the energy efficiency labels. Here we take a look at some of these labelling schemes and other ways to ensure your purchase is as green as possible.

Office equipment

Watch out for the Energy Star Level, which will rate your purchase on how energy efficient it is. Not only will they reduce the impact on the environment, but it will also save you money on your energy bills. This is a voluntary scheme, so you should bear in mind that if a new computer doesn't carry the Energy Star logo, it could still be a green option.

Similar to the Energy Star scheme is the TCO Certification Scheme, a worldwide environmental rating scheme that covers notebooks, desktops, projectors, phone headsets and visual display units. IT equipment is assessed under a range of environmental and ergonomic criteria to help consumers select a product that will have as little impact on the environment as possible. The difference is that while there are various different levels of accreditation, such as TCO07 and TCO Certified Edge, within each level there is no room for grading – a product will either be awarded the certification or if it fails to meet stringent criteria it won't be awarded.

White goods

'Under EU law most white goods, such as washing machines, freezers and fridges, are required to display an energy efficiency level. This label will inform you just how energy efficient your new appliance will be.'

Under EU law most white goods, such as washing machines, freezers and fridges, are required to display an energy efficiency level. This label will inform you just how energy efficient your new appliance will be. The system uses a nine-grade system from G (the least efficient) right up to A++. The A+ and A++ grades are more recently introduced to keep up with the evolving energy efficient technology, while products that have been graded as E, F or G have been banned from being sold since 1998.

According to the European Household Appliance Association, the average European manufacturer of domestic appliances has been able to gain a 20% increase in efficiency every four years – demonstrating just how important eco-friendly products are to consumers. To reflect this, a new labelling system for fridges and freezers will be coming into force in 2011, with other appliances adopting new labelling methods at a later date. With this move, the current A+ and A++ grades will be replaced with A-20%, A-40% and so on, to indicate how much more efficient they are than a standard A grade appliance.

Cars

In the past, you may have assessed your new car purchase based on performance or reputation. However, as green issues become ever more important, taking into account the eco credentials of a new car will become just as relevant. This will help you pick a model that will have the least impact on the environment and could even save you money. As the Act on CO_2 campaign states: 'If everyone buying a brand new car chose the most fuel efficient car in its class, CO_2 emissions from new cars could be reduced by up to 24% and save up to three months' worth of fuel per year.'

If you're purchasing a brand new model within Europe, it will be sold with information giving you the CO_2 emissions and fuel economy of the vehicle. Since 2005, cars have been categorised into bands from A to F according to the carbon emissions emitted per kilometre. To achieve an A rating, the car must release less than 100g of carbon per km, while less eco-friendly cars will be graded F (over 186g of carbon per km). In 2006, a G rating was added, and two years later a further six bands were added, so all cars are now rated A (the best) through to M (the worst at over 255g of carbon per km).

Of course not everyone will be in a position to purchase a new car – but this doesn't mean having to sacrifice your green intentions. In 2009, a used car label was rolled out to dealerships. The label is colour-coded and provides easy-to-understand details for customers about the car's make and model, its CO_2 emissions, the estimated fuel cost over 12,000 miles and MPG (miles per gallon). Although the scheme is a voluntary initiative developed by the Low Carbon Vehicle Partnership, there are currently over 2,000 dealers participating in it.

Ten green cars

Manufacturers are developing ever-greener cars, making the purchasing decision for consumers that little bit easier. According to What Green Car, some of the greenest cars to look out for are:

Peugeot iOn electric

Nissan LEAF electric

Toyota Plug-in Prius hybrid

Audi A1

Toyota Auris hybrid

Citroen DS3 HDi

Volvo S60

Ford C-MAX

Honda CR-Z hybrid

BMW 5 Series

Labels to look for

The Leaf Marque

The LEAF Marque is awarded to farm products that meet the LEAF farming principles, such as sustainable farming and environmental responsibility. Look out for it on fruit and vegetables.

The Marine Stewardship Council

The Marine Stewardship Council logo is all about sustainable and well-managed fisheries. Look out for it on seafood products.

Organic

The organic status is defined by law – so you can be confident that if a product carries an organic label (such as the Soil Association, DEMETER or Organic Farmers & Growers), it will be carefully assessed as containing no chemicals such as pesticides or genetically modified ingredients. Look out for it on everything.

Energy Saving Trust Recommended

A scheme run by the Energy Saving Trust, this logo will ensure that your purchase is one of the most energy efficient on the market. Look out for it on home appliances, lighting, electricals, heating products, computing equipment, glazing and insulation.

PEFC Council (Programme for the Endorsement of Forest Certification Schemes)

This logo will confirm that products have been independently audited to prove they come from sustainably managed forests. A similar scheme is that of the Forest Stewardship Council. Look out for it on timber products, including paper.

Rainforest Alliance

The Rainforest Alliance works closely with famers and foresters to ensure that they are operating in an environmentally and socially responsible way. Look out for it on paper, bananas, tea, cocoa, coffee and timber.

Oeko-Tex Standard 100

This is an international certification scheme for the textile industry. Products bearing this label will have been tested independently for a range of harmful substances, defined by the latest scientific reports and legal regulations. Look out for it on garments.

European Ecolabel

Managed by DEFRA, the Europe-wide Ecolabel will reassure you that products have minimal impact on the environment at all stages of the product's life cycle. Look out for it on cleaning products, toilet tissue, clothing and tourist accommodation.

Going green grants

Because going green makes so much sense, there are a lot of grants available for people to make green changes to their home.

Warm Front Scheme

In England, the Warm Front Scheme (known as Warm Homes in Northern Ireland, Energy Assistance Package in Scotland and the Home Energy Efficiency Scheme in Wales) provides up to £3,500 to households to improve their heating and energy efficiency. The grants do have some eligibility criteria (such as being in receipt of certain benefits or being over a certain age), but full details can be found by visiting the websites listed in the help list.

The Sustainable Development Fund

The Sustainable Development Fund is run by National Parks and is funded by money from DEFRA and the Welsh Assembly Government. The grant scheme is all about encouraging people to find sustainable ways of living and in England projects can receive up to 75% of their total cost (up to 50% in Wales). In the past, National Parks has supported projects such as wood burning systems and rainwater harvesting through to eco buildings and training 'green ambassadors'.

Energy Saving Trust

By simply typing in your postcode and answering a few questions, you can quickly find out what grants and schemes you're eligible for – covering everything from solar heating through to loft insulation. Its comprehensive database holds details for hundreds of grants from a range of providers – government schemes, energy supplier initiatives, retailers' and installers' promotions.

Energy suppliers

Because energy suppliers are increasingly coming under pressure to achieve targets designed to improve energy efficiency, many suppliers provide a range of offers to help you improve on this. All UK suppliers who have a certain number of customers are obliged to attain these targets under the Carbon Emissions Reduction Target (CERT) scheme so you, the customer, can benefit. Plus it doesn't matter who supplies your energy as you can still apply for a grant. Start your search by checking out each of the main suppliers' schemes and see what you're entitled to.

'Going green doesn't have to involve vast costs, but it makes sense that when the time comes to replace your boiler, washing machine, car or other expensive item you consider its environmental impact.'

Pay As You Save (PAYS)

Although this is currently a pilot scheme, you can still benefit from this Energy Saving Trust scheme. The finance solution enables households to invest in energy efficient projects and micro-generation technologies (such as solar panels) with no upfront costs. Instead, you can make repayments over a longer period, which are lower than the predicted energy bill savings. The Department of Energy and Climate Change (DECC) has made available £4 million worth of funding for pilot schemes in Birmingham, London Borough of Sutton, Sunderland, Stroud, Surrey and Sussex. Visit the Energy Saving Trust website for more information.

Plus

To help you lead a greener life, look out for scrappage schemes on everything from cars and windows through to boilers. It's also worth contacting your local authority to find out if they can provide any grants for local residents in order to install energy efficient measures in the home, such as loft insulation.

Summing Up

Going green doesn't have to involve vast costs, but it makes sense that when the time comes to replace your boiler, washing machine, car or other expensive item you consider its environmental impact. Although green technology still comes with a slightly higher price tag, the potential cost savings will soon mount up. So investing in the short term could lead to savings in the long term.

There are several logos to look out for when shopping to be sure you can identify products that truly do offer an eco benefit. Don't be afraid to ask questions of the people selling the item you're interested in.

In addition, do some research to find out what loans or grants you might be eligible for which could enable you to install solar panels or double glazing or to help you make green home improvements such as loft insulation. As everyone is waking up to the idea that we need to do more and the government is setting industry targets, more companies are offering incentives to ensure they're doing their bit.

Help List

Act On CO2

www.actonco2.org.uk
A cross-government initiative from the Department of Energy and Climate Change (DECC), the Department for Transport (DfT) and Department for Environment, Food and Rural Affairs (DEFRA). The website offers information and ideas for greener living and a carbon calculator.

BareEscentuals

www.bareescentuals.co.uk
A brand of make-up and beauty products which are 100% pure and use natural ingredients.

Bio D

www.biodegradable.biz
Manufacturer of environmentally responsible, ethically sound, hypoallergenic household cleaners.

Blackle

www.blackle.com
Energy saving search engine, powered by Google but with a black screen to save energy.

British Assessment Bureau

www.british-assessment.co.uk
The British Assessment Bureau is a specialist in ISO Certification and bespoke assessment within the UK, including the environmental certification ISO 14001. That standard was first published in 1996 and today identifies those companies that are committed to making changes to be more environmentally sound.

Burt's Bees

www.burtsbees.co.uk

Natural and environmentally-friendly cosmetic and skincare products.

Carplus

www.carplus.org.uk
Carplus is a national resource for people looking into sustainable transport information. On the charity's website you can find details and information of car share clubs that are already setting up, as well as information about the benefits of joining and how to set your own scheme up.

Community Repaint

www.communityrepaint.org.uk
An award-winning UK network of 65 community-based paint reuse projects. Consumers and businesses can donate unused or half-used tins of paint to the project, which then distributes them to community and voluntary groups or individuals in need.

Compassion in World Farming

www.ciwf.org.uk
Organisation campaigning to end cruel factory farming practices.

DEMETER

www.demeter.net
A certification given to produce that is wholly organic.

Department of Energy and Climate Change (DECC)

www.decc.gov.uk
Government department responsible for climate change and energy policy in the UK.

Department for Environment, Food and Rural Affairs (DEFRA)

www.defra.gov.uk
Government department responsible for policy and regulations on the environment, food and rural affairs.

Earthborn Paints

www.earthbornpaints.co.uk

Earthborn is a range of eco paints and natural varnishes designed to provide a healthier and environmentally friendly alternative to conventional paints and varnishes.

E-Cloths

www.e-cloth.com
A range of cloths which only require water to clean greasy and dirty surfaces. Using only water eliminates the need for cleaning products.

Ecolibrium Paints

www.ecolibriumpaints.com
Ecolibrium is an award-winning range of natural paints and garden furniture treatments made with over 90% natural ingredients.

Ecos Organic Paints

www.ecospaints.com
Range of solvent-free paints and varnishes.

Ecover

www.ecover.com
Eco-friendly cleaning product range.

Empty Homes Agency

www.emptyhomes.com
An independent charity helping people create homes from empty properties and campaigning for more empty homes to be brought into use for those with housing needs.

Energy Assistance Package (Scotland)

www.energysavingtrust.org.uk/scotland
In Scotland, the Energy Assistance Package can help fund energy efficient home improvements. More information about the package can be found on the Energy Saving Trust's Scottish website alongside advice for householders, businesses and communities.

Energy Saving Trust

www.energysavingtrust.org.uk
The Energy Saving Trust is a non-profit organisation that provides free and impartial advice on how to stop wasting energy. The site is packed with news, details on available grants and schemes and plenty of ideas to become more energy efficient.

Energy Star Scheme

www.energystar.gov
Energy Star is a joint scheme from the US Environmental Protection Agency and the US Department of Energy to help consumers preserve energy, protect the environment and save money. The Energy Star logo is awarded to energy-efficient products and the website includes a search function to find products awarded the logo.

Enhanced Capital Allowance Scheme

www.eca.gov.uk
This scheme enables businesses to claim 100% first year capital allowances on their spending on green machinery – this includes energy-saving plant and machinery, low CO_2 emission cars, natural gas and hydrogen refuelling infrastructure and water conservation plant and machinery.

Fleet Department

www.fleetdepartment.co.uk
Fleet Department is a cooperative organisation aiming to provide businesses with up to 250 vehicles, with the cost savings, dynamics and high levels of administrative support previously only enjoyed by the very large corporations.

Forest Stewardship Council

www.fsc-uk.org
The Forest Stewardship Council is an international organisation that is committed to promoting responsible management of the world's forests. It provides a certification scheme for forestry and forest products, which is searchable on its website.

Freecycle

www.uk.freecycle.org
A website where you can give or receive no longer needed items for free.

Go Real

www.goreal.org.uk
The Real Nappy Information Service provides support and advice for parents looking to use real nappies and details of their nearest nappy provider or laundering service.

Good News for Polar Bears

www.goodnewsforpolarbears.org
A recycling website where members can give or receive no longer needed items for free.

Green Buying

www.greenbuying.co.uk
This informative site provides busy buyers with a one-stop shop for green purchases. As well as a green services directory, you can also find an eco shop selling green office supplies and the scrap shop – a free-to-use trade waste exchange. The Sustainable Purchasing Healthcheck is also useful for any office looking to become more sustainable.

Green Electricity Marketplace

www.greenelectricity.org
This website lists and analyses all of the available green energy tariffs to help promote renewable energy. The website has a handy postcode search to find out what's available in your area and information on the various types of green energy.

Green Office Week

www.greenofficeweek.eu
A campaign dedicated to help green-over offices. On the website there's lots of support and encouragement for workers looking to go green and the opportunity to take part in webinars on the topic. The campaign takes place once a year and is support by Avery.

Green People

www.greenpeople.co.uk
Green People is a forward-thinking company which specialises in organic body care products.

Home Energy Efficiency Scheme (Wales)

www.heeswales.co.uk
In Wales, the home energy efficiency scheme can provide help for homeowners to make improvements to their homes to enhance energy efficiency. The website has full details of what help is available and the eligibility criteria.

LEAF

www.leafuk.org
LEAF is committed to a viable agriculture which is environmentally and socially acceptable and ensures the continuous supply of wholesome, affordable food while conserving and enhancing the fabric and wildlife of the countryside for future generations. LEAF is also responsible for administering the LEAF Marque certification scheme.

Liftshare

www.liftshare.com
Car sharing website where you can enter your regular route and pair up with people doing the same or similar journey.

London Cycle Hire Scheme

www.tfl.gov.uk
Administered by Transport for London, the cycle hire scheme enables those in the capital to get around on bike. The website lists docking information so you know where to pick up a bike from.

Love Food Hate Waste

www.lovefoodhatewaste.com
A campaign group to reduce the amount of food wasted in British homes. The website has lots of information, recipe ideas, storage information to prolong the life of food and a portion size planner.

Low Carbon Vehicle Partnership (Low CVP)

www.lowcvp.org.uk
The Low CVP is a partnership of over 350 organisations from the automotive and fuel industries, the environmental sector, road user groups, government and more. It is an action and advisory group to promote the reduction of carbon emissions through the use of alternative fuels.

Marine Stewardship Council (MSC)

www.msc.org
A global organisation working with fisheries, seafood companies, scientists, conservation groups and the public to promote the best environmental choice in seafood.

Method

www.methodproducts.co.uk
Eco-friendly cleaning products.

Moltex Oko

www.moltex.de
Manufacturers of a range of eco-disposable nappies – the site is available in German and English.

Morgan Lovell

www.morganlovell.co.uk
This office interior design, fit out, refurbishment and business relocation company works with BREEAM assessors and LEED designers to help customers create a more sustainable and environmentally friendly workplace.

National Car Share

www.nationalcarshare.co.uk
Car sharing website where you can enter your regular route and pair up with people doing the same or similar journey.

Natural House

www.natural-house.co.uk
Soil Association-certified organic cleaning product range.

Natural Organic Edible

www.noecosmetics.com
Range of skincare products derived from 100% natural, organic and edible ingredients.

Naturebotts

www.naturebotts.co.uk
An online shop with everything parents need for their babies. Stocks a huge range of eco-disposable nappies, real nappy supplies, clothes, shoes, books, toys, feeding equipment and first aid products. Products are natural and kinder to the environment than conventional ranges.

Neal's Yard Remedies

www.nealsyardremedies.com
Organic and natural range of health and beauty products.

Nicky Amos CSR Services Ltd

www.nicky-amos.co.uk
A consultancy that works with organisations to develop, communicate and report on sustainable business practices and strategy.

No More Post

www.nomorepost.com
Online paperless postal system which allows businesses to send post to an online postal database that customers can then access and view via one safe and secure online account that will always remain spam free.

Nothing Nasty

www.nothingnasty.com
Range of 100% natural skincare products.

Oeko-tex

www.oeko-tex.com

The Oeko-Tex® Standard 100 is a certification scheme that provides the textile industry with a global, uniform standard for assessing the presence of harmful substances. A product carrying this certification will have been independently tested to confirm it is free from certain substances deemed unsafe.

Ofgem

www.ofgem.gov.uk

Ofgem regulates the electricity and gas markets in Great Britain.

On-Pack Recycling Label

www.onpackrecyclinglabel.org.uk

Launched by the British Retail Consortium, the On-Pack Recycling Label scheme aims to deliver a simpler, UK-wide, consistent, recycling message on both retailer private label and brand-owner packaging to help consumers recycle more material, more often.

Organic Farmers and Growers

www.organicfarmers.org.uk

Organic Farmers and Growers Ltd is one of a number of control bodies accredited by DEFRA and is approved to inspect organic production and processing in the UK.

Origins

www.origins.co.uk

Range of natural skincare products.

PEFC

www.pefc.co.uk

Any product carrying the PEFC logo has come from a sustainably managed forest.

Pesticide Action Network UK

www.pan-uk.org

PAN UK is an independent, not-for-profit organisation that works to eliminate the dangers of toxic pesticides to improve the environment we live and work in. The website includes a pesticide database and links to over 5,000 articles, reports and books relating to the environmental and health impact of pesticides.

Pots of Paint

www.potsofpaint.com

Range of natural, non-toxic paints.

Rail Europe

www.raileurope.co.uk

Specialists in tickets and passes for rail travel around UK. The website offers comprehensive details of routes, prices, journey information, trips to take and more.

Rainforest Alliance

www.rainforest-alliance.org

Works closely with farmers and foresters to ensure that they are operating in an environmentally and socially responsible way.

Recycle Now

www.recyclenow.com

A useful site packed full of tips, advice and information on recycling. You can find your nearest recycling bank or input what it is you're looking to recycle to find the nearest collection point. There are also lots of resources to make the most of your waste.

Responsible Travel

www.responsibletravel.com

A travel agent dedicated to responsible holidaying. The website has details of over 300 tour operators, 600 accommodation providers and a handy search function to find the perfect holiday for you and your family. You can also find unedited holiday reviews.

Soil Association

www.soilassociation.org
A charity campaigning for planet-friendly, organic food and farming. The Soil Association is also one of the bodies that can certify organic produce according to strict guidelines.

Stagecoach

www.stagecoachbus.com
The UK's most common bus company, currently Stagecoach is investing in more environmentally-friendly technologies.

Streetcar

www.streetcar.co.uk
A self-service, pay-as-you-go car scheme to ensure you always have transport when you need it but without the environmental impact of owning your own car.

Sustainable Development Fund

www.nationalparks.gov.uk
Run by National Parks, the sustainable development fund is a grant scheme to support projects with economic, social and environmental benefits. National Parks' role is to conserve and enhance the natural beauty, cultural heritage and wildlife in areas of countryside and promote learning opportunities to the public.

TCO Certification Scheme

www.tcodevelopment.com
The TCO certification scheme enables customers to select IT and office equipment that has benefits to the environment as well as for the end user. The website includes a TCO certified product search function and advice on product selection.

Warm Front Grant (England)

www.warmfront.co.uk

The Warm Front scheme helps to fund home improvements to enhance energy efficiency. The scheme provides a package of insulation and heating improvements up to the value of £3,500 (£6,000 where oil, low carbon or renewable technologies are recommended).

Warm Homes Grant (Northern Ireland)

www.warm-homes.com
Funded by the Department for Social Development, this Northern Ireland scheme enables eligible households to receive financial help for insulation and heating improvements.

Weleda

www.weleda.co.uk
Natural skincare and beauty range.

What Green Car

www.whatgreencar.com
A useful tool when looking to purchase a new car. The website has information on green cars, the latest news from the industry, a function to work out how green your current model is and reviews of the eco-friendly cars available.

Women's Environmental Network

www.wen.org.uk
Women's Environmental Network is a charity dedicated to educating, informing and empowering people to care for the environment. The website has lots of information, useful guides and fact sheets.

World Health Organisation (WHO)

www.who.int
WHO is the directing and coordinating authority for health within the United Nations system. It is responsible for providing leadership on global health matters, shaping the health research agenda, setting norms and standards, articulating evidence-based policy options, providing technical support to countries and monitoring and assessing health trends.

Zipcar

www.zipcar.co.uk
A car sharing scheme enabling people to ditch costly ownership of cars and instead join a scheme and only use a car when they need it.

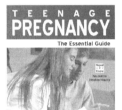